WILD HUNTERS

❖ ❖ ❖

PREDATORS IN PERIL

*To Holly
good luck in your
work with Mountain
lions*

MONTE HUMMEL
SHERRY PETTIGREW

❖ ❖ ❖

Illustrations by
ROBERT BATEMAN

*Robert Bateman
Nov/94*

with John Murray

ROBERTS RINEHART
PUBLISHERS

Published in the United States of America by
Roberts Rinehart Publishers
121 Second Avenue
Niwot, Colorado 80544

International Standard Book Number 1-879373-27-0
Library of Congress Catalog Card Number 92-60265

Printed in the United States of America

CONTENTS

To Evan Jack
for a lifetime of memories upcountry

And to Joyce and Ian McTaggart-Cowan
for their lifetime contribution
to conservation

ACKNOWLEDGEMENTS

This special U.S. edition of *Wild Hunters* is based on World Wildlife Fund's (WWF) *Conservation Strategy for Large Carnivores in Canada*, which was released in November 1990. Melanie Watt, who helped draft the strategy, originally suggested that an edition for the general reader was needed. We are indebted to the many other people who provided input for that original document, in particular the seven-person Scientific Steering Committee comprised of Dr. Fred Bunnell, Dr. Lu Carbyn, Dr. Stephen Herrero, Martin Jalkotzy, Dr. Ian McTaggart-Cowan, Dr. Ian Stirling, and Dr. John Theberge. Mr. John Murray from the University of Alaska provided the introduction to the U.S. edition and the section on A Conservation Strategy for Large Carnivores in the United States.

In connection with WWF's Large Carnivore Strategy, special acknowledgement is due to Dr. Herrero, as well as to Alistair Bath and Harold Dueck, two graduate students at the University of Calgary, for preparing an excellent 1988 background working paper, *Carnivore Conservation Areas*, funded by a grant from WWF. Harold Dueck also authored a subsequent report, *Carnivore Conservation and Inter-agency Cooperation: A Proposal for the Canadian Rockies*, prepared for WWF and the Canadian Parks Service in 1990. We fully acknowledge these studies as the sources for most of what appears on the subject of minimum viable populations and Carnivore Conservation Areas in this book.

ACKNOWLEDGEMENTS

Robert Bateman provided a personal foreword for which we are very grateful. Within his demanding schedule, he also found the time to create the magnificent new sketches in our book and donated them to us for this use. In preparing these drawings, Bob used a new technique which he intends to explore further in the future—partly India ink drawn with a twig, a quill pen, spattering, airbrushing, scraping off and dissolving with alcohol! Bob's images and his generosity speak for his outstanding personal commitment to conserving the wild living things that inspire his work. Thank you very much, Bob.

Interviews were conducted with a number of people who kindly provided professional insight, personal anecdotes, and advice on what we should include. They are: Vivian Banci, Fred Bunnell, Lu Carbyn, Peter Clarkson, Dick Dekker, Graham Forbes, Gordon Haber, Hank Halliday, Alton Harestad, Stephen Herrero, Maurice Hornocker, Fred Hovey, Martin Jalkotzy, Charles Jonkel, Paul Joslin, Rosemarie and Pat Keough, George Kolenosky, Wayne McCrory, Bruce McLellan, Ian McTaggart-Cowan, François Messier, Malcolm Ramsay, Joe Robertson, Ian Ross, Candace Savage, Barbara Scott, Ian Stirling, John Theberge, Jay Tischendorf, Dennis Voigt, John Weaver, and Rob Wielgus. We thank you all.

We are also grateful to the Ontario Ministry of Natural Resources for allowing us to use information from the species range maps that appear in their 1987 book *Wild Furbearer Management and Conservation in North America,* and to all the other publishers and researchers who gave us permission to reprint from their work. Dick Dekker published an excellent book entitled *Wild Hunters* in 1985, which documents his adventures over twenty-five years with wolves, foxes, eagles, and falcons in western Canada. He graciously allowed us to use the phrase "wild hunters" in our book title as well.

Thanks are due to Jan Marsh, who typed our first text so that we could revise and edit from the original version, and to Jill and Basil Seaton and to Glen Davis, who each reviewed the entire manuscript and provided many helpful comments. We are grateful to Ian Stirling, Bruce McLellan, George Kolenosky, John Theberge, Martin Jalkotzy, and Vivian Banci who examined our species chapters for technical and scientific accuracy. And we thank Laurie Coulter, our editor, for doing such an excellent job of editing the manuscript, chickenpox and floods notwithstanding!

Finally, we extend our sincere appreciation to Fredrik S. Eaton, who provided a grant to WWF to cover the costs associated with researching and preparing this book.

Every book has a source of inspiration. For us, that source is the North American wilderness. We acknowledge the resourceful wild bears, wolves, cougars, and wolverines that are out there even as you read these words. May they always be.

MONTE HUMMEL AND SHERRY PETTIGREW

FOREWORD

From our perch at the end of the twentieth century, with an overload of information, we humans can survey the history and geography of our planet and contemplate the impact we have had on our global home. We have never been better informed or more powerful. Now, at last, we must take full responsibility for our actions. The twentieth century has been a "dynamite" century in every sense of the word . . . flashy, explosive, and very destructive. The last fifty years in particular have seen more annihilation of our natural and human heritage than any other period in history. In earlier years we could perhaps use the excuse that "we knew not what we did." We can use that excuse no longer. We now know that we are doing. During this decade, we must decide on the kind of world we want to leave to future generations. We are playing God whether we want to or not.

One of the most conspicuous decisions we must make very soon is about large predators living wild and free. Since the days of early cave dwellers, we have had a powerful and competitive relationship with these animals. Sometimes the cave bears pushed us from our caves; sometimes lions and tigers drove us from our freshly killed quarry. At other times, the tables were turned. As we acquired weapons and made more use of our wits, *Homo sapiens* began to win the majority of conflicts. As the competition became more one-sided, we came up with a new approach. We called it sport. Now predators which were no threat or competition were sought out and slaughtered for reasons of *machismo* and fun. But the greatest threat has been the spread of agriculture and industry. The organization of plants and animals for our use leaves no room for large predators. Often, there is not even room for their prey species. When this happens, of course, wolves may prey on calves and cougars may prey on lambs. They have no choice. When hikers and campers invade the territory of a bear or a lion, *Homo sapiens* may become prey species themselves.

How can large predators continue to survive wild and free? There is only one answer: These large animals need large spaces. Fortunately, some of these spaces still exist in North America, many of them with predator populations in place. However, almost all of them are threatened by industrial-scale forestry or megaprojects in energy and mining. There is no place for megaprojects in the precious lots of wilderness which remain at the end of the twentieth century. We now must take responsibility and show that we are civilized enough to share this planet with our ancient worthy opponents.

ROBERT BATEMAN

INTRODUCTION TO THE U. S. EDITION

by John Murray

When I was a boy growing up in southern Ohio in the 1950s, I often wondered what it would be like to live in an area with large predators. The only sizeable carnivore in our neighborhood was a racoon that occupied a buckeye woodlot behind the old Quaker church, and it disappeared when the new subdivision went in. As more and more development took place, my friends and I began to range farther from our homes. Finally one year we bicycled all the way to the Ohio River, where we spent the afternoon looking for fossils on a limestone outcrop overlooking the wide valley. The view to the west commanded our attention, and after awhile we just sat down and stared at the horizon. Somewhere beyond the cornfields, cow pastures, and far oak hills were the Rocky Mountains, a place, we'd heard, where bears, wolves, and wild cats were still as common as they once had been in Ohio. We knew instinctively that those distant green solitudes were as vital to the health of our country as our hearts were to the life of our bodies.

The opportunity to move west arose when I was seventeen. My father, who worked for the Environmental Protection Agency (EPA), was transferred from Cincinnati to Denver. That spring the two of us packed into a high basin and camped beside a cirque lake. It was on that first evening, gathering wood among the subalpine firs, that we found a set of fresh black bear tracks on a lingering snowbank. We built the fire a little higher that night and slept a little less soundly, but that was fine, because when we returned to the city we knew we had been in bear country. All kinds of arguments—religious, ethical, aes-

thetic, economic, biological—can be made for preserving predators, but in the end, it comes down to moments like that, a father
and son camped beneath Ursa Major with a wild bear somewhere on the frosty slopes.

The United States has a diverse climate and topography. As a
consequence, its large predators occupy a complex array of habitats. In the extreme north of Alaska, for example, arctic tundra
covers the land surface and pack ice joins the continental coast
early each winter; polar bears, grizzly bears, arctic wolves, and
wolverines live in this lightly-human-populated region. Forty-
five degrees to the south, on the other hand, we find tropical
hardwood forests and fringing coral reefs in south Florida similar to those of the Caribbean oceanic islands and the Central
and South American littorals. Black bears and endangered Florida panthers inhabit the Everglades, as do American alligators
and American crocodiles; historically red wolves were also present throughout the peninsula. Between the two geographic extremes of Alaska and Florida are many unique predator
habitats—hot Sonoran deserts, cold intermontane deserts, semi-
desert chaparral, southern hardwood forests, and southern
swamp and marshland—habitats that do not exist in Canada.
Although the biology of predators north and south of the
international border is generally similar, as is the history of
Euramerican persecution, the current status of the species is
often quite different. The wolverine, for example, although considered "vulnerable" in western Canada, is either "rare" or "endangered" at the southern limit of its North American habitat in
California, Utah, and Colorado. Similarly, seventeen subspecies
of the gray wolf are currently distributed across the breadth of
Canada, whereas the gray wolf has long been extinct from most
regions in the United States and is now found only in remnant
populations in northern Montana, northern Minnesota, northwestern Wisconsin, and the Upper Peninsula of Michigan.
There are several other significant differences between Canada and the United States *vis-á-vis* large predators. For one
thing, Canada has a smaller human population (27 million
compared to 270 million) but more wilderness parks and physi-
cally larger parks than the United States. Also, Canada has a
larger number of national parks in its eastern provinces; most of

the parks and conservation areas of the United States are concentrated in the far western part of the country, especially in the Rocky Mountain region and the Pacific Northwest. The parks of the United States, and by inference the resident predators, are threatened by much more intensive use than those of Canada. For example, in the six-state Rocky Mountain region, there were 29.6 million visitors in 1991, an increase of 7 percent over the 27.7 million visitors of 1990. Glen Canyon National Recreation Area in Arizona and Utah was the most heavily visited area in the region in 1991, with 3.1 million recreation days, an increase of 107,000 over the previous year. Following Glen Canyon were Yellowstone National Park in with 2.9 million visitors (an increase of 97,000 over 1990), and Rocky Mountain National Park with 2.7 million visitors (an increase of 104,000 over 1990). Nationally, over 267 million National Park Service visitations were registered in 1991, this from a population ten times that of Canada. Clearly, what is done, or not done, in the next decade with respect to the conservation of predators in the United States will influence the relationship between humankind and nature living in this country for generations to come.

Polar Bear
Two polar bear populations exist in arctic Alaska. A 1988 book by Ian Stirling estimated a total of at least 2,000 polar bears in the northern Alaska population and in the western Alaska population. Although polar bears are protected by the Marine Mammal Protection Act of 1972 and the Endangered Species Act of 1973, regulations currently permit Alaska's Inupiat people to hunt polar bears for subsistence purposes. There are no limits or restrictions on sex, age, or method of hunting. The sole stipulation is that there be no waste of body parts. Conservationists and others have challenged the appropriateness of such practices in areas where a species may be scarce or otherwise endangered by resource development projects.

The immediate threat to the polar bears of Alaska is the proposed development of oil and gas resources on and near the coastal plain of the Arctic National Wildlife Refuge (ANWR) in northeastern Alaska. Polar bear maternity dens have been confirmed near Barter Island, at the delta of the Hulahula River, near the Sadlerochit River, on Carter Creek, on the Niguanak

River, near the Egaskrak River, and on the Kimikpaurauk River. All of these areas would be adversely effected by the energy development project. Many scientists believe that the possibility of disrupting these sensitive and critical habitat areas is so great that the development of the ANWR should be forever banned; the U.S. Congress is supporting their position and has banned all development for the time being.

Another threat to the polar bears is the cyclical warming and cooling trends that have been recorded in the Arctic. If the climate becomes cooler, denning areas could be extended deeper into the river drainages of the ANWR. Conversely, if the climate becomes warmer, fewer females could reach favorable denning areas. This could negatively effect natality for the species not only in Alaska but throughout the circumpolar north. Only time will tell if in fact—as many believe—increased levels of carbon dioxide in the atmosphere from industrial pollution is causing a global warming phenomenon.

Grizzly Bear

For many, the consummate North American predator is the grizzly bear. No other predator so dominates the frontier literature of the American West or so powerfully evokes the idea of the North American wilderness. It was only recently in the United States that societal values began to change with respect to this great bear, and measures were taken to ensure its protection.

Currently the grizzly is, with several important exceptions, maintaining much of its historic range in Alaska. Experts believe there are at least 32,000 grizzlies in the state. However, there is cause for concern in several local situations. For example, a proposed mining development and a proposed salmon fishery seriously threaten the brown bears of the McNeil River Bear Sanctuary, an internationally known bear observation area on the Alaskan peninsula. Visitors to this location often observe up to 60 bears feeding on the salmon at the McNeil falls. Farther north, the grizzlies of Denali National Park, which is approximately the size of Massachusetts, are imperiled by increased tourist development in the western Kantishna region, by potential mining developments, and by a meteoric rise in the number of visitors using the park. Maintaining current automobile re-

strictions on the Denali park road in the future will be essential
to preserving grizzly bear viewing opportunities in the park;
over 90 percent of all park visitors see a grizzly bear in the wild.
Elsewhere in Alaska, as in neighboring British Columbia, clearcutting of coastal rain forests has seriously damaged both grizzly
and black bear habitat. Planned development of oil and gas
resources on the coastal plain of the Arctic National Wildlife
Refuge (ANWR) in northeastern Alaska could also seriously
impact prime grizzly feeding sites. Caribou is an important prey
species for these bears, and the Porcupine caribou herd uses this
plain for its calving grounds. Hunters are permitted to harvest
one grizzly every four years. At this time, controlled hunting is
not perceived to be a danger to the species; it was uncontrolled
hunting coupled with habitat loss that ensured the extinction of
the species in the American West.

Only a handful of grizzly populations, a total of perhaps 1,000
animals, remains in the lower forty-eight states. Yellowstone
National Park is home to about 200 grizzlies; between 440 to 680
grizzlies live in the Northern Continental Divide region; the
North Cascades grizzly population was only confirmed in 1991
and there are no population estimates available; as of 1989, there
were 19 grizzlies wearing active radio collars in the Selkirk
mountains; and in the Bitterroot and San Juan mountains no
population estimates are available. The grizzly bear is categorized as either "threatened" or "endangered" in Wyoming,
Idaho, Montana, Washington, and Colorado, in accordance with
the 1973 Endangered Species Act. The U.S. Fish and Wildlife
Service administers all plans designed to help the grizzly populations recover.

The state of the grizzly bear in Yellowstone National Park has
been a subject of controversy and concern since the Craighead
study concluded abruptly in 1971 on the eve of the park's centen-
nial. The key to the health of this isolated population, which
was decimated in 1971-1973 following the dump closures, is the
number of breeding females. Much of the scientific debate about
the future of the bears centers around one critical fact—unless
natality and recruitment exceed mortality the species is doomed
in the park. Supporters of current management policy believe
there are a sufficiently high number of breeding females and
that the species is on the road to recovery. Critics, however,

assert that the number of breeding females is dangerously low and that the bears could become extinct in Yellowstone. Discussion also invariably focuses upon Grant Village and Fishing Bridge, two controversial tourist developments that are located in what almost all scientific authorities agree is critical grizzly bear habitat.

Yellowstone is not the best grizzly habitat. It is certainly not as rich an environment as the northern Rockies or the southern Rockies. The park contains a high and relatively dry mountain plateau that was—until the forest fires of 1988—heavily forested and suffering from the effects of a burgeoning elk population, which was formerly held in check by wolves (now extinct in the park). The forest fires opened up previously timbered country and have provided the secondary growth—particularly berry bushes—so prized by the grizzly bears. This will definitely help the bears in the 1990s.

No discussion of the grizzly in the lower forty-eight states would be complete without mentioning the situation in the San Juan Mountains of southwestern Colorado. Although unconfirmed sightings and reports of grizzly bear activity persisted well after 1952, when a grizzly bear was killed in the mountains, by the 1970s state and federal officials had lost hope that the last relict population of southwestern grizzlies still existed. Suddenly, and much to the surprise of the scientific world, an adult female grizzly bear was killed by a hunter in the San Juans on September 23, 1979. A postmortem examination of the carcass indicated the grizzly had probably borne cubs in the past, indicating the presence of a male and cubs in the area. The state of Colorado undertook a two-year study of the area in which the bear was killed and found recent dig sites and a recent grizzly den. In 1982 an outfitter reported seeing a blond grizzly sow and two cubs in the same vicinity, and researchers later discovered evidence that the bear had indeed been in the area. Despite these sightings and reports, federal and state land and wildlife managers have failed to take the necessary steps to protect grizzlies in the region. Black bear hunting (using bait as a lure) and livestock grazing, both sources of historic conflict with the grizzly, continue. In March 1992, state wildlife officials confirmed that hairs found in the south San Juans were grizzly bear hairs. Because the grizzly is a federally listed endangered species in Colorado, many conservationists hope that the federal government will begin a more aggressive campaign to save this precious population of grizzlies including reintroductions to build the bear population.

Black Bear

Black bears occupy a diversity of habitats in the United States. In the northeast, the primary habitat types consist of beech-maple forests and red spruce-balsam fir forests. Although there are not large tracts of federal wild lands in this part of the country, several timber companies own vast areas—particularly in Maine—which have been managed as *de facto* wildlife preserves for several generations. Black bear hunting through leases is also a major enterprise conducted here by the landowners. Black bears in the southeast occupy two major types of habitat: the oak-hickory woodlands of the southern Appalachians and the pine parkland and cypress swamps of the coastal lowlands. Additionally, an endangered isolated population of black bears is found in the Everglades of south Florida. More federal land is available in the southeast than in the northeast, particularly in the mountains of Tennessee and North Carolina, and in places—such as Great Smoky Mountains National Park—the species is abundant. Quaking aspen and spruce-fir forests, interspersed with open parks and high mountain bogs, predominate in the Rockies, where the black bear is common at all elevations. Farther west, equally good habitat is found in the Sierra Nevadas and Pacific Northwest, where the black bears can exploit a rich berry and mast crop at lower elevations, and then move into higher forests of redwood, hemlock and Douglas fir. The species is either rare or endangered in Florida, Massachusetts, South Carolina, Georgia, Texas, Louisiana, Alabama, Kentucky, Ohio, and New Jersey. The greatest challenge facing researchers in these and other areas where the black bear is scarce is acquiring reliable census data.

The black bear is behaviorally more adapted to forest environments than to open country habitats. Consequently, it thrives in areas where the grizzly has now become extinct, such as in New Mexico, Arizona, and California. In some areas, however, the black bear has become so numerous and so adapted to human beings that problems have arisen. This is particularly the case in

Great Smoky Mountains National Park and in Yosemite National Park, where incidents occur each year involving campers or hikers and black bears. Like the polar bear and the grizzly bear, the black bear can be a very dangerous animal in particular situations; for example, when an individual bear is surprised at close distance or when a mother perceives a threat to her offspring.

In addition to habitat loss, other major threats to black bears in the United States include overhunting and underestimation of black bear mortality losses from such difficult-to-quantify sources as poaching and natural factors. A recent seven-year study of black bears in Gunnison National Forest in Colorado by bear biologist Tom Beck concluded that, even in an area closed to bear hunting, the population numbers are lower than had been previously thought, and that deaths from poaching and from other causes are higher. Beck recommended a much more conservative approach to bear hunting policy, including the possible elimination of spring bear hunting, which can, and does, remove females (and their immature cubs) from the population. His recommendations could probably be applied to the other three black bear habitat regions of the United States.

In 1992, the black bear was listed in Appendix II of the Convention on International Trade in Endangered Species of Flora and Fauna (CITES) to which the U.S. is a signatory. This was because of threats to the black bear through international trade in its parts, especially gall bladders and paws, but it will require more careful record keeping on state populations and management practices in the future.

Wolf

Like the mountain lion, the wolf—which in North America included the gray wolf, red wolf, and Mexican wolf—was once widely distributed across the whole of the United States. Wolves greeted the earliest Spanish explorers in Florida and New Mexico, and were recorded by British and French explorers and pioneers in New England and New France, as well as along the Pacific Coast. As a result of decades of persecution, in the United States, the gray wolf is today found only in northern Minnesota, northwestern Wisconsin, the upper peninsula of Michigan, and northern Montana. Unconfirmed reports of wolves occasionally come from North Dakota, Idaho, Arizona, and New Mexico.

In 1986, the federal government proposed that the gray wolf be restored to three areas in the Rocky Mountains: northwestern Montana in and around Glacier National Park (this has now occurred naturally through repopulation across the international border); northwestern Wyoming in and around Yellowstone National Park; and central Idaho in and around the Selway-Bitterroot Wilderness Area. Because the natural colonization of these last two areas is unlikely, plans have been made to translocate wolves into them. The 1982 amendments to the 1973 Endangered Species Act contain provisions for the reintroduction of endangered or threatened species into areas they formerly inhabited. These populations are referred to as "experimental populations," existing when "the population is wholly separated geographically from non-experimental populations of the same species." Major opposition has naturally come from cattle ranchers and woolgrowers, but it appears at this writing that wolf recovery may soon proceed in Yellowstone. This will involve approximately ten pairs of mated wolves which will be released at a number of locations in the backcountry where prey species—primarily elk, mule deer, bison, antelope, and mountain sheep—are numerous. Should this recovery proceed successfully, it is possible that a relocation effort in Idaho may take place. Park managers in Rocky Mountain National Park have also called for the restoration of wolves to their park in order to control the burgeoning elk populations, which they have no legal method of managing.

The red wolf once inhabited the entire Old South, from central Texas and central Oklahoma to the Atlantic, and from the Gulf of Mexico north into the lower Midwest. Apparently there was little overlap between the gray wolf, which inhabited all other areas of the United States, and the smaller and more specialized red wolf. The red wolf appears in the wild as a smaller version of the gray wolf, with shorter, ruddy-colored fur, and a lighter bone structure. By 1970, the red wolf was found in only a few areas in southwestern Louisiana and northeastern Texas along the Gulf Coast. A number of these individuals were captured, checked for their genetic purity (red wolves will interbreed with coyotes), and then placed in about half a dozen captive breeding programs. Since 1988, red wolves have been released at five different locations: Bulls Island, South Carolina; Horn Island, Mississippi; St. Vincent Island, Florida; Alligator River National Wildlife Refuge, North Carolina; and Great Smoky Mountain National Park, Cades Cove, Tennessee. Because in each case the support of local residents has been sought and reinforced through public education programs, these projects have had the societal acceptance necessary to be undertaken successfully. These popular wolf-release programs have no doubt influenced politicians and special interest groups in the Rocky Mountain region, who are now taking a more serious look at gray wolf reintroduction there.

Historically, the Mexican wolf was distributed through central, south, and west Texas; southern New Mexico; southern Arizona; and throughout the Sierra Madre mountains and associated lowlands of northern Mexico. Scientists believe these wolves still persist in Mexico. Following passage of the Endangered Species Act in 1973, federal authorities made an effort to collect as many live specimens of the endangered Mexican wolf as they could; currently there are about forty Mexican wolves in captivity. In 1979, a Mexican Wolf Recovery Team was appointed to consult with the U.S. Fish and Wildlife Service about potential release sites for the captive wolves. Their final recommendations included several locations in west Texas, southern New Mexico, and southern Arizona. Two locations now seem likely: Big Bend National Park in Texas and/or the White Sands Missile Range in New Mexico. Although the Defense Department initially raised objections to the White Sands Missile Range as a potential release site, these objections were withdrawn in 1990. Strong opposition still remains in both Texas and New Mexico from various livestock associations. Another possibility would be the 900-square-mile Gila Wilderness Area in southwestern New Mexico. The Gila offers an array of habitats and traditional prey species—including white-tailed deer and javelina—for the Mexican wolves. Additionally, many of the grazing allotments in the Gila have been purchased by state wildlife managers for wildlife use, so there is less of the historic conflict with livestock in this area than previously. Because the recovery plan for the Mexican wolf is now in place, it seems likely that its reintroduction will move forward in the 1990s, as will those of the red wolf and gray wolf.

Cougar

At the time of Columbus's arrival in the New World in 1492 the cougar probably had the most extensive range of any carnivore in the Americas. It was likely absent from only a few inhospitable areas in the United States, such as Death Valley or the barren regions of the Dakota badlands. Today the species is found only in the Florida Everglades and in the following western states: Montana, Wyoming, Colorado, New Mexico, Texas, Arizona, Utah, Idaho, Nevada, California, Oregon, and Washington. In these regions, the cougar lives in close proximity to its major prey species, the white-tailed deer and the mule deer.

In common with the black bear, the two major threats to the cougar consist of habitat loss—particularly near urban areas where lion populations have benefited from hunting bans—and overharvesting through hunting. In contrast with bears and wolves, however, the cougar breeds at any time of the year, and females can be killed by sport hunters after they have been separated from their young and treed by dogs. Proper education of guides, outfitters, and hunters can reduce the number of these unnecessary killings and ensure a sustained-yield harvest consistent with preservation of this predator.

As in eastern Canada, yearly reports of cougar sightings in the eastern United States encourage belief that the species persists there. Reliable sightings have been confirmed in the central and southern Appalachian Mountains, the Ozark Mountains of Arkansas and Missouri, the Wichita Mountains of Oklahoma, southeastern Arkansas, and northern Louisiana. Cougars were killed in eastern Tennessee near Pikeville in 1971 and in northwestern Pennsylvania in 1967.

An isolated and critically endangered population of cougars is found in south Florida. Approximately 30 to 50 of these animals are believed to inhabit Big Cypress National Preserve and Everglades National Park west of Miami. A serious problem for this isolated population is the lack of genetic diversity. Inbreeding has resulted in a loss of 70 percent of the genetic diversity originally present in these wild cats. Scientists have recently noticed a number of crippling defects: 70 percent of the newborn males have only one testicle, up to 90 percent of the sperm is nonfertile or abnormal, and some lions are now born with major heart-valve anomalies. In 1990, state authorities began a captive breeding program to save the Florida cougars, but in 1991 the Fund for Animals sued to block these measures. At this writing, a temporary agreement has been reached permitting the capture of six kittens. The cougar is Florida's state animal and it is to be hoped the species can survive inbreeding as well as the threats presented by cattle grazing, water diversion projects, citrus production, residential land development, and road building.

Wolverines
Outside Alaska the wolverine is found in only a few areas in the United States: northeastern Minnesota, western Montana, northern Idaho, northwestern Wyoming, central Colorado, northeastern Utah, and in the high mountains of Washington, Oregon, and northern California. In most of these areas it is considered by wildlife managers to be either rare or endangered. In Colorado, for example, a group of researchers studied 265 reports of wolverine sightings and concluded that only 18 were probable sightings, including three imported wolverines that escaped from a zoo, a fresh wolverine skull found in a remote mountain region, a photograph of a wolverine on a snow field, a photograph of a female and three cubs in the mountains, and two wolverines that were released by a wildlife film production company near Aspen. Additionally, a wolverine killed in March 1979 in Utah a few miles west of the Colorado state line was thought to have come from Colorado.

The species is found throughout Alaska, although it is reported to be rare on the arctic coastal plain and in some tundra and marsh regions of western Alaska. Throughout its domain in the lower forty-eight states, the wolverine remains an elusive and little-understood carnivore.

Living in Alaska as I now do, my perspective on U.S. predators is somewhat similar to that of Canadians. Like them, I am separated geographically, if not politically, from the lower forty-eight states, and live in a region less ravaged by unbridled development and overpopulation than the contiguous United

States. Having lived in both the lower forty-eight and Alaska, I have a comparative view of the challenges facing carnivores. On the one hand, I have seen how terribly nature has been treated by the ever-expanding human civilization; on the other hand, I am presented with a vision here of what nature was like in pre-Columbian times, and of what it could be again in the future.

Wherever civilization has gone, predators have suffered. They are always the first vertebrate species to disappear. Afterwards come nostalgic and guilt-ridden attempts at environmental repair. We must, though, whatever the material cost, ensure that these precious resources and predators—as much a part of our freedom as any word in the Constitution—are preserved for posterity.

Not long ago I was hiking, as I do for many weeks each summer, in a remote valley of Denali National Park. I had not gone very far down the worn caribou trail, stopping to pick some seasoned blueberries and examine a shed moose antler that a porcupine had gnawed, when a willow ptarmigan, still in its summer brown, flushed from an alder patch just ahead. As it flapped heavily downhill, flying with so much effort it would seem easier to walk, I was amazed to see a wolf about 9 meters (30 ft.) away. The wolf looked at me impassively and then turned away into the alders.

After an appropriate wait, I walked to where the wolf had been standing. It was nowhere to be seen. From its former vantage point, I had a view of the entire headwater valley. There was a snowstorm in progress on the glacier—a preview of the blizzard to come—and a wall of sleet was pouring over the high alpine ridge to the west. I spotted some caribou down on the river, coming my way, and I took out my binoculars and brought them closer. There were five bulls, ranging in size from a spindly four-pointer to an old fellow with double shovels and so many tines I lost count. Two of the younger bulls broke from the others and galloped up the hill toward me. The large hanging flaps of white breast fur that squared the front quarters swayed loosely as they slowed to a canter. They stopped and briefly locked antlers, then trotted off together, their racks held high over their shoulders.

The wall of sleet steadily engulfed the caribou and, one by one, they dissolved into the storm. It was then, scanning the slopes, that I saw the wolf again, far below. It was loping along, hidden by an intervening ridge, on an intercept course with the caribou. It moved rhythmically with visible strength, head lifted, tongue hanging out, and tail up. This big, powerful-looking male carried a weight that indicated success as a hunter. And then I saw the other wolves—three of them closing rapidly on the other side of the herd. At that moment I lost view of the wolves, as I had the caribou, to the wall of bad weather. I stood there for a long time on the side of the hill, trying to decide whether to hike deeper into the valley, and risk disturbing the hunt, or to remain on the overlook as the storm obscured the view, and forever wonder at the outcome.

It is my hope and prayer that such moments, which elevate and define our lives as much as anything else, will forever be accessible to the citizens of the United States. The wolf, the polar bear, the grizzly, the black bear, the cougar, the wolverine, and the jaguar are an affirmation to us that our country is still young and strong. To commune with them is to taste the natural freedom of which Rousseau and Locke wrote so eloquently. Of all the social contracts we have, perhaps none is more vital to the security of our future than that which we have with nature, and with those symbols of wilderness, the wild predators.

1.
Why CONSERVE TOP PREDATORS?

Space is air for the great beasts who roam the earth.
Now is their final breath.
JOHN WEAVER, MONTANA WOLF RESEARCHER

IT IS IRONIC THAT THE MOST POWERFUL WILDLIFE SPECIES, PRESIDING, AS they do, at the top of the food chain, have consistently fared the worst in relation to people.

To understand the tragic scope of what we have done, imagine for a moment that you have a green pencil in your hand and a black and white map of the Northern Hemisphere before you. Imagine also that you could color in the areas occupied by the big cats, bears, wolves, and other wild, land-based predators 200 years ago when the human population was only 10 percent of what it is today. Now imagine an eraser capable of rubbing out your careful green shadings. This powerful eraser represents the recent impact of human civilization.

As the eraser goes to work, it wears away the edges of your green areas: the range of large predators shrinks with the spread of settlement and agriculture. Because it happens slowly and is almost unnoticeable from year to year, its effects are measured in decades. In North America and Eurasia, this shrinkage moves inexorably from the south toward the north. *The grizzly bear disappears from 99 percent of its range in the American West.*

Sometimes the eraser rubs out a spot in the middle of a green area, then spreads the white patch out from there: such is the impact of cities. *The wolf disappears from fourteen countries in central Europe, from 95 percent of the lower United States, and from the southern portions of Canada.*

In places, a large green area is completely erased: here, the top predators no longer exist at all. They have become extinct — forever lost from that part of the world. *The North American prairies now show up as an immense blank band through Canada and the United States, emptied of plains bison, bears, and wolves.*

Sometimes the eraser leaves a faint green hue: in these areas top predators are still found, but in very low numbers, and are in constant danger of extinction. *The wolverine and cougar maintain a ghost-like, endangered presence in eastern North America.*

In some regions, a few green spots are left behind, places the eraser couldn't reach: remnant predator populations are protected in national parks or wildlife refuges. *Grizzlies in the United States are hanging on in just a few protected areas, most of which are shared with Canada.*

Finally, where top predators are still abundant, the map bravely holds its green. However, these areas are also in danger of being erased. *The black bear faces the threat of poaching for international trade in its teeth, claws, paws and gall bladder.*

The pattern varies across the map, but the overall picture is strikingly obvious. As the eraser continues, the green-penciled areas steadily become smaller and smaller. And, when you look more closely, you realize that, apart from the Soviet Union, the last remaining green-covered areas of any size in the Northern Hemisphere are found in Alaska, and in Canada, especially in the West and North. Therefore, one thing is clear: if any green remains on the map, if there are going to be still-wild populations of top predators, it will be in those areas that many of them will take their last stand.

This bird's-eye view of what has happened to large carnivores provides the first powerful argument for concerted new efforts to conserve them. If we don't make such efforts now, there's good reason to believe that history will repeat itself, and these animals will ultimately be lost. Many experts support work by World Wildlife Fund, and other groups, to highlight the need for large-carnivore conservation and the need to set aside conservation areas right now. As Montana wildlife biologist John Weaver says, "It's going to be difficult to implement because, in Canada, you're not in a crisis yet. But, in my view, history tells us that the handwriting's on the wall."

This potential loss is one that many of us care about deeply. American conservationist Aldo Leopold called us "those who can't live without wild things." And what about future generations who might choose to "live with wild things"? We should not rob them of the choice as to whether or not they will want these charismatic "superspecies" to be part of their world. We have no right to foreclose their options; if we do, we must be prepared to hear them say they wish we *had* cared.

<h2>HANG-UPS ABOUT "PREDATORS"</h2>

To understand the role of predators in natural systems, the first thing most people will have to do is get rid of some language baggage. The word "predator," used in a human context, brings with it all kinds of nasty connotations.

"Predator" is derived from the Latin word for "prey" (*praeda*), which refers to "booty" or a "victim of a harmful or hostile person." In this book, we're talking about carnivorous or "meat-eating" predators. The word "carnivorous" is derived from the Latin words for "flesh" (*caro carnis*) and "to eat" or "devour" (*vorare*). Dictionaries describe the behavior of predators as "plundering, pillaging, robbing" or "preying on others." Today the word "predator" is used in the business world to describe corporations engaged in clandestine or openly aggressive take-overs, and in the entertainment world to convey a sense of exciting (and mindless) violence in television, video, and movie fare. Put all these definitions together, and you have the popular image of a predator — "a pillaging, plundering, devouring, hostile, harmful, voracious, flesh-eater"!

In contrast to this loaded view, you might think that scientists, in their cool, objective world, would have a tidy definition of "predator" on which they all agree. Not so. Some ecologists use an all-encompassing definition of "predation," which is essentially "one living thing being eaten by another." Whether what is being eaten lives or dies isn't important. For example, they would call deer feeding on plants "predators," although the plant in the ground lives on. According to this definition, even parasites feeding on substances inside the deer would be called "predators." However, for the purposes of this book, we use a less general definition: a predator is: "any living thing that kills and eats another for food."

By this definition, of course, humans are predators as well. But our level of predation, through direct killing and habitat destruction, bears little resemblance to any natural process. In his book,

Bear Attacks: Their Causes and Avoidance, Stephen Herrero ironically observes, "If a book entitled *People Attacks* were written for bears, it could only depict our species as being typically blood-thirsty killers, aggressive, dangerous, often inflicting fatal injury to bears."

Even our simplified definition of "predator" leaves the field wide open for the inclusion of not only humans, but also other mammals, birds, fish, reptiles, amphibians, insects, spiders, and even insect-eating plants. For this book, we have selected the six North American "large" carnivores or "top" predators: the polar bear, grizzly bear, black bear, wolf, cougar, and wolverine. They were chosen not because the smaller carnivorous mammals such as the fisher, badger, lynx, and bobcat are doing fine, but because, within any group of animals, it is generally the larger species that are eliminated first. The plight of these six top predators may well serve as an indicator of what can happen to many of the smaller predators as well.

At this point, a word about "killing" might also be useful. In most societies, killing another person is considered morally wrong. Furthermore, death is usually feared, ritualized, and even "overcome" through religious beliefs. In many respects, then, we deny death or hide it away. Often these human concepts about death are ascribed to wildlife through childhood stories, movies, or cartoons that portray animals as people. "The Big Bad Wolf," for example, is seen as a murderer, threatening everything from the Three Little Pigs to Little Red Riding Hood. Barbara Scott, who conducted wolf research on Vancouver Island, doesn't mince words regarding her feelings about these stereotypes, especially those that always portray predators as threatening or bad. She would "destroy all children's propaganda on wolves. I'd get rid of all those fairy tales. Those stories even got *me* wondering whether I was going to go into the field or not. It made me realize how strong that influence is in your childhood years. I'm going through some of that with my own boys right now."

At the other extreme are Yogi Bear and Gentle Ben, friendly caricatures of grizzly bears that suggest that predators pose no danger at all.

Given our childhood exposure to these stereotypes, it is little wonder that many of us bring human concepts to our understanding of predators and predation, or killing, in natural systems. We do this by imparting human values to the act of killing in nature. We judge it to be "good" or "bad," whereas it is simply a fact of life. In reality, death is a precondition for life in the wild. The death of a deer, for example, is necessary for the life of a wolf to continue. Similarly, the act of killing, repeated over and over many times in nature, is a process necessary for the perpetuation of a complex biological life system. Yet, although there is no such thing as a "good" or "bad" wolf in nature, human values often cause people to "feel sorry" for a deer killed by a wolf, especially if it is a "baby" deer. Or, some people become incensed if all of a deer isn't eaten by a wolf. They talk about wolves "wantonly killing more than they can eat," and "wolf-lovers" are told they would change their tune if they "could see what a wolf does to a deer" — the suggestion being that the wolf is cruel and wasteful in the way it kills. However, George Kolenosky, who works for the Ontario government researching wolves and black bears, replies, "They're simply trying to make a living, the same as you and I, and to raise their young. They generally have to kill something else in order to make that living. Wolves can't go to the supermarket and buy food off a shelf, like we can; they have to go and find it."

When a wolf kills livestock, such as sheep or cattle, it is seen to have overstepped the bounds of legitimate predation, because it has gone outside the natural system to kill something that is raised and directly valued by humans. Then the wolf is persecuted, even though we were the ones who plunked our farms down in its habitat. And when a bear attacks or kills a human, the bounds are seen to have been violated again, even though the attack took place in the bear's domain. In this case, the killing of one of our fellow humans is considered a "morally wrong" act committed by the bear. The animal, of course, is found and usually killed in return. Interestingly, many of these feelings and opinions are voiced in reverse when people argue about whether it is right or wrong for us to kill or "prey on" predators.

The point is not to judge whether or not these strongly held views are justified, but simply to demonstrate that, because predators by definition "kill and eat" to survive, they are the focus of confused and emotionally explosive opinions. These kinds of viewpoints are inevitable. They vary from person to person, but we all have them in one form or another, and they can stand in the way of understanding the real role of predators in natural systems.

THE ROLE OF PREDATORS IN NATURAL SYSTEMS

One of the most frequently heard arguments in favor of conserving top predators is that they play an important role in natural systems. The argument usually goes like this: as long as predators are present, the predator and prey populations, for example, wolves and deer, naturally fluctuate in relation to each other. These fluctuations occur within a long-term "natural balance" or "equilibrium." The argument continues, that, without predators, prey numbers grow almost unchecked until the prey population overbrowses its habitat, then the prey numbers crash as a result of disease or starvation. The moral of the story? Leave predators in the system to play their natural and obviously important role.

However, over the past seventy-five years, there has been a great deal of debate about the precise nature of the effects predators have on prey populations. In this debate, particular emphasis has been put on the question of how important predation is in relation to other factors that can affect prey populations — factors such as climate, disease, or food supply. Why, for example, should the wolf be "blamed" for reducing deer numbers if something else, such as harsh weather or disease, actually caused the decline?

The early view was that predators "regulate" or control prey numbers, thereby causing population fluctuations. It was generally believed that, as predator numbers increase, prey numbers decrease, until the system reaches a point where the reverse occurs: predator numbers decrease and prey increase. The population-cycle theory of predation was largely based on fieldwork in the 1920s carried out on small mammals by a famous English ecologist, C.S. Elton.

This view of predation was refined in the 1930s and 1940s after an equally famous study on mule deer in the Kaibab Plateau of Arizona by Aldo Leopold. Here, wolves, coyotes, and cougars had been deliberately killed off, primarily by ranchers. The deer population soared from 4,000 to 100,000 in twenty years, then crashed to below 30,000 in the next twenty years. These results are frequently cited to support the view that prey populations will "erupt" in the absence of predators, overbrowse their habitat, and subsequently "crash" in numbers. However, later work showed that the absence of predators was only one factor causing the Kaibab deer eruption. It was discovered that, in the aftermath of fires, the vegetation had changed, providing an increased food supply for the deer. In addition, the decline in numbers of domestic sheep and cattle had reduced competition for the area's food supply, allowing deer numbers to increase.

After these observations were made, researchers throughout the 1960s argued over which factor — predation, food supply, climate, disease, competition within a prey population — had the greatest effect on prey numbers. Not surprisingly, the debaters eventually discovered that all of these variables affect wildlife numbers, but that such variables as hard winters, good summers, and fires (through their effect on habitat) are the ultimate causes of fluctuation in wildlife populations.

To further confound the issue, natural factors that have negative effects on some wildlife species can actually be positive for others, as Vivian Banci discovered in her study of wolverines in the Yukon: "In a harsh winter, there are a lot of caribou and moose dying. This means the wolverine are doing great because there are lots of carcasses to scavenge. So that year's recruitment of young (the number of young wolverine that survive) would be high. When there's a really mild winter, the caribou and moose are doing fine, but now the wolverine can't find anything to eat."

Research into how different variables affect wildlife numbers led to further investigation into the relationship between predation and the "ultimate" factors, or ecological limits of natural systems. For example, it was suggested that predator and prey populations fluctuate in relation to a "nutrient/climate ceiling." In simple terms, this means that the nutrients present in the ecosystem ultimately limit everything else because they affect the growth of plants, which, in turn, affects food supply, prey numbers, and ultimately predators. Climate, through temperature and rainfall, sets a similar limit on food chains. It is important to note that these factors affect ecosystem processes and wildlife numbers whether predators are present in the system or not.

The current view is that predators may not cause population fluctuations of prey, but they can hasten declines in prey numbers caused by other factors, or they can lengthen the periods when prey are at lower numbers. These effects occur in tandem with, and are limited by, broader environmental factors such as the nutrient/climate ceiling. A famous 1970s field study by Lloyd Keith, an ecologist at the University of Wisconsin, supports this view. It indicated how lynx and hare populations fluctuate in relation to each other, as well as in relation to factors such as weather and snowfall occurring in the boreal forest.

The work of Elton, Leopold, and Keith still serves as a guidepost for scientific thinking on this subject. However, new theories suggesting that ecosystem behavior is essentially chaotic and that there may be no such thing as stability or equilibrium in natural systems are throwing into question even those long-standing theories and models. In the end, it is unlikely that scientists will ever adequately understand something as complex as the interaction between predators and prey, and how such relationships are affected by the many other factors operating in nature. Such is particularly the case in "multi-predator/prey" systems where there are several predator species, for example, wolves, bears, and cougars, interacting with several prey species, such as moose, caribou, and mountain sheep and goats.

Although we may never know exactly what the relationship is between predators and prey, we can be sure of one thing: the predator/prey relationship is an important one and should be respected. Therefore, the role of predators in natural systems stands as a solid reason for their conservation.

PREDATORS AND PROTECTED AREAS

Since predators occupy the top of the food chain, their presence can tell us something about the health of the natural system itself. If top predators are present and healthy, chances are that the ecological processes "below" them are also in good shape. In this sense, they serve as indicators of the overall well-being, or integrity, of the system. Their home range, or habitat requirements, are also generally the largest of any members of the natural system. A quick, "back of the envelope" way to calculate how large a protected area or wilderness park must be to represent and maintain the integrity of a particular ecosystem is that it should be big enough to accommodate a viable population of its top predators. By accommodating them, we likely accommodate everything else. Of course, that is not going to be strictly true in every case because some species or elements of an ecosystem may not always be included by simply making a protected area large enough to conserve top predators. Furthermore, it is difficult to determine what a viable population of these animals is. Nevertheless, this rule of thumb is generally helpful in determining what size a protected area should be.

The maintenance of ecological integrity is required by policy in the Canadian National Parks Act and by the U.S. Forest Service. Indeed, the ability to conserve top predators could be used as one measure of how well these two countries are doing in meeting this policy goal; if we are not saving top predators, we're not saving true wilderness. And if we are not saving true wilderness, we will not save top predators.

Predators and Biological Diversity

"Biological diversity," a somewhat technical-sounding term, has recently become fashionable. It refers simply to the richness of life on earth — the full diversity of both species and ecosystems.

When conservationists talk about "preserving genetic diversity," they are usually referring to maintaining the variety of species on our planet, in their natural habitats. Ecologist and WWF Gold Medal winner Paul Ehrlich has likened species to rivets in an aircraft. Unfortunately, the rivets are popping out of Spaceship Earth, which can lose only so many before the threshold number is exceeded and a crash occurs. We don't know what that threshold number is. Obviously, if we are to save the spaceship, we must retain the rivets. However, the reverse is also true. If we care for and save the entire spaceship as a system, we avoid losing the rivets. By conserving entire ecosystems, we retain and save species.

Currently, we are losing species from the earth at a frightening rate. Up to one million micro-organism, insect, plant, bird, and mammal species now living could become endangered or lost by the year 2000. By conserving top predators, many of which are already formally listed as endangered, we will be contributing to the urgent worldwide task of preserving biological diversity. By failing to conserve biological diversity, we jeopardize all life on earth.

Species loss, or extinction, is a biological fact of life, contributing to the larger process of evolution. In fact, probably 90 percent of the species ever found on the earth have passed into extinction.

However, these have been replaced by new species that have evolved to maintain the planet's overall biological diversity. The current diversity of species will also shift over time, as species are lost, added, or changed.

In this context, predators are worth conserving because of their role in the evolution of life-forms. Predators have a number of evolutionary effects. They prevent some wildlife species, such as deer and elk, from building up to such high numbers that they literally eat themselves out of house and home. Predators are also a mechanism for natural selection. They coevolve with their prey species, and affect the physical attributes of their prey, which, in turn, affect the predators' own attributes. A faster deer thus produces a faster wolf, which produces a bigger, stronger deer, and so on. Big wolves have evolved in conjunction with big prey. However, the little red deer on the Isle of Man in the Irish Sea may have evolved because predators disappeared. In the absence of predators to flee or fight off, greater strength and size were no longer required in the deer. In this long-term, evolutionary sense, then, predators ensure survival of the best-adapted. It has even been suggested that the way the wolf hunts has evolved in a "prudent" manner that insures its own food supply. By usually taking the most easily caught prey — older and and very young animals — the wolf leaves the prime, reproductive part of the population to supply more "wolf food" for the future.

Conserving predators so they can play their rightful evolutionary role is one way to help insure the long-term survival of our global life-system. It would also be a long-overdue gesture of respect by our own species for the natural evolution of other species.

PREDATORS AND PEOPLE

Predators should be conserved not only for their role in natural systems, but also for their importance in the religion, legends, and everyday lives of aboriginal peoples.

The polar bear and the Inuit serve as just one example. In the legends of these Arctic marine people, the Goddess of the Sea was regarded as supreme, but the second most powerful spirit was that of the polar bear. This spirit often became the guardian or protector of an important person. It could also shift form, from human to animal, an aspect reflected in Inuit sculptures and prints depicting beings that are half human/half bear. A skinned-out bear carcass looks very human, which may account, in part, for the close identification of people with polar bears.

To the Inuit, polar bears were given, not taken. After a polar bear was killed, sufficient time had to go by before another one was killed to allow the soul of the first bear to return to its family. The bear's soul warranted respect, and it was capable of being offended if proper rituals weren't followed. It would not give itself up to undeserving individuals. Bear hides were used for clothing and sleeping robes, the meat was used for human and dog food, and its teeth were carved into ornaments and amulets.

Today, this magnificent animal continues to play an important role in Inuit life. Native artists strive to represent the white bear's spirit in paintings, prints, and sculptures. The hides and hunts provide an important cash income for Arctic hunters who also make their traditional knowledge and skills available to help scientists learn more about these bears.

In contrast to that of aboriginal cultures, modern civilization's relationship with predators has been disastrous. It is a sad story of hatred, fear, misunderstanding, and persecution through bounties, poisoning, shooting, trapping, and habitat destruction, and has led to many local and some nationwide extinctions. Many large-carnivore specialists have noted that agriculture, in particular, and top predators don't seem to mix. In fact, there is a direct link between the most developed agricultural lands in Canada and the loss of large carnivores in these areas. Wolves and grizzlies have been extinguished from the Canadian prairies, and now we're pushing grizzlies from the foothills of Alberta and the British Columbia interior grasslands — areas sought after for ranching and farming.

Clearly, then, the "importance" of top predators to most non-native people over much of the past 200 years has been a negative one, based on the perception of these animals as vermin. In his introduction to a 1909 book titled *Wolf and Coyote Trapping*, C.J. Harding wrote, "There are no animals that destroy so much stock as wolves and coyotes, as they largely live upon the property of farmers, settlers, and ranchmen to which they add game as they can get it. While these animals are trapped, shot, hunted with dogs, etc., the day of their extermination is, no doubt, far in the distance."

Though such attitudes die hard, and some landowners still do have legitimate problems with losses of livestock to predators, in recent years overwhelming public support has developed for the intelligent conservation of wildlife such as the grizzly bear and the wolf. These species have now become inspiring symbols of wilderness — a natural world that is fast disappearing. There is no doubt that these new attitudes will be the driving force behind the conservation of large carnivores, *if* such attitudes can be informed and nourished soon enough, and *if* they are effectively brought to bear on decision-makers in government and industry.

PREDATORS, TRAPPING, AND HUNTING

All the top predators in this book are legally classified as either "game animals" or "furbearers." That means they can be killed by hunters and trappers, subject to laws and regulations that are supposed to insure that the number of animals killed is within safe conservation limits. Obviously, these animals are important to people who hunt and trap them.

On the one hand, that is good news. Hunters and trappers are often in the forefront of conservation, particularly in their support of measures to protect the habitat of these animals. And that is very important, since habitat loss or degradation is the most fundamental reason for wildlife decline.

On the other hand, hunters and trappers seem to want to have these species around so that there are enough to kill. Some people find this motive objectionable in principle, even if the level of killing is sustainable and not a threat to the wildlife population as a whole. Problems also arise when hunters and trappers want to kill wildlife in or around protected areas such as parks, or when hunters advocate predator-control programs in order to increase what, to them, are more desirable animal species to hunt, such as moose and deer.

It is not the purpose of this book to present arguments for or against hunting and trapping. World Wildlife Fund, for example, acknowledges hunting and trapping as a fact. WWF's primary concern is to make sure that such activities are carried out within safe conservation limits. Such a position may take into account the biological effects of hunting and trapping, but what about the ethics involved?

First, it should be recognized that the decision to accept only sustainable use of wildlife *is* an ethical commitment — a commitment to maintain wildlife species and populations for the long term. Without this commitment, there will be nothing for people on either side of the use/non-use debate to fight about. And, certainly, many wildlife populations are already in danger of being lost forever. In this context, squabbling over whether individual animals should be "used" or "not used" while, for example, large-scale loss of habitat is causing serious threats to entire wildlife populations is essentially fiddling while Rome burns.

Second, it should be recognized by those advocating hunting and trapping that many top predators are particularly vulnerable to such activities. As they have evolved to occupy the top of the food chain, there are not as many of them in wildlife systems as there are other species that serve as prey in those systems. Most predators' reproductive characteristics reflect that fact. Many, for example, have very low reproductive rates, don't become sexually mature until they are relatively old, have small litters, and may not reproduce at all in years of poor food supply. Their large home-range requirements mean they are found in low numbers, thinly spread out over enormous areas. For these reasons, shooting a top predator such as a grizzly bear is quite different from killing a rabbit or a mule deer, which has evolved as prey. Since predators are special by virtue of their role in natural systems and their reproductive characteristics, these facts must be reflected in both the regulations governing and the attitudes taken to the exploitation of these mammals, if indeed they are going to be exploited. In some areas of Canada, these facts are beginning to be acknowledged. On the whole, however, we still have a long way to go.

Finally, there are several other ethical concerns, which are not necessarily anti-hunting but which relate to the type of hunting practiced in relation to large carnivores. For example, poaching or illegal killing, a form of hunting, should be opposed by hunters and non-hunters alike for both biological and ethical reasons. Although poaching is generally opposed, many people feel they can "do nothing about it," that "it's too big a problem," or that the person doing the poaching "is my friend." Large numbers of Canada's top predators, and many other wildlife species, are illegally killed each year as a result of this public hesitancy and conspiracy of silence.

Another ethical concern relates to the reasons why hunting is practiced. Based on the end-use to which wildlife is put, many people differentiate between sport hunting for meat, sport hunting to obtain a trophy, and sport hunting simply "for fun"; or between sport and subsistence hunting. Many hunters themselves object to the commercialization of wildlife, through game farming or through businesses that rely on guiding non-resident hunters on successful hunts, because it has never been easy to insure that the profit motive operates within ecological limits. In fact, it rarely does.

Some "high-tech" hunting innovations are also objectionable. One example is the use of dogs wearing radio-transmitting collars so the hunter can more efficiently locate the dogs and the bear or cougar being chased. Another is "will-call" hunts for cougars where a hunter doesn't even buy a hunting license until a cougar has been chased by hounds and treed. At that point, the guide telephones the waiting hunter, who then buys a license, shows up, and shoots the wild cat out of the tree.

In other cases, it is what hunting brings with it that causes problems. For example, building roads and all-terrain-vehicle or snow-mobile trails in wilderness backcountry may cause more disruption to wildlife than the act of hunting itself. But the damage was brought about in the first place by the request for hunting.

These are only some of the ethical and biological questions that can and should be raised about hunting. Hunting has become a minority activity in Canada, practiced by less than 10 percent of the public, and that percentage appears to be getting smaller year by year. While non-hunters shouldn't trample on the rights of a minority, neither should hunters as a minority command a disproportionate share of wildlife-agency budgets. No doubt, the best approach is to fashion alliances between conservation groups and "user groups" on fundamental issues such as habitat protection. In this way, more energy could be spent working together on areas of mutual concern, rather than fighting while the house comes down around us.

PREDATORS FOR THEIR OWN SAKE

Philosophers delight in reminding us that, even by thinking about something, we are relating it to ourselves. We cannot escape the fact that we sense and understand the world through human perceptions linked with human emotions. Nevertheless, a strong argument can be made that natural systems, including predators, are worth conserving for their own sake, completely independent of their importance or usefulness to people.

John Livingston, in his challenging book, *The Fallacy of Wildlife Conservation*, defines wildlife conservation as "the preservation of wildlife forms and groups of forms in perpetuity, for their own sakes, irrespective of any connotation of present or future use." He argues that any concept of conservation motivated by human use, even spiritual and artistic uses, is self-defeating and, in the long run, detrimental to wildlife.

Although Livingston's position might seem extreme to some people, he is not alone in wanting to conserve wildlife for its own sake. In describing the wolf's "right to exist," the Manifesto on Wolf Conservation (see Appendix A), drafted by the World Conservation Union (formerly the International Union for the Conservation of Nature and Natural Resources) emphasizes that "this right is in no way related to their [the wolves'] known value to mankind."

When asked what changes he would make if he could, regarding the conservation of large carnivores, Fred Bunnell, a wildlife/forestry professor from the University of British Columbia, said, "If I could do it with a magic wand, I'd create a sense of intrinsic worth of all species among humans. That would be number one, because if we could get there, then we could get past a lot of the rest of it!" Wildlife experiences similar to the following ones he describes have led many of us to the same conclusion: "It's the delight taken just watching a mother bear with her cubs. Or the things that give you a jolt — a black bear wandering through the woods, coming up to a log that looks reasonably stable and — 'whomp!' — ripping it apart with a few blows and gobbling up the grubs. Or the incredible, lithe grace of a cougar. Or the fierce and almost obnoxious determination of wolverines!"

Aldo Leopold was another supporter of the principle of conserving wildlife for its own sake. In 1949, in *A Sand County Almanac*, he stated, "It is only in recent years that we hear the argument that predators are members of the community, and that no special interest has the right to exterminate them for the sake of a benefit, real or fancied, to itself. Unfortunately this enlightened view is still in the talk stage." Ironically, Leopold began as an advocate of getting rid of predators to protect and produce more big-game animals for human use. But in his essay "Thinking Like a Mountain," he concluded that "only the mountain has lived long enough to listen objectively to the howl of the wolf." The moment that changed his view, and indeed his entire life, came while he was watching the "fierce green fire" die in the eyes of a wolf he had shot. "I was young then, and full of trigger-itch; I thought that because fewer wolves meant more deer, that no wolves would mean hunters' paradise. But after seeing the green fire die, I sensed that neither the wolf nor the mountain agreed with such a view."

Of course, we can't "think like a mountain" because we are not mountains, but humans. Leopold reminds us, however, that certain meanings are hidden from us *because* we are humans, including "the meaning in the howl of the wolf, long known among mountains, but seldom perceived among men." If we *could* "think like a mountain," perhaps we could truly understand the importance of conserving top predators.

PREDATORS' CURRENT STATUS IN CANADA

The most common question asked about the six species dealt with in this book is, "Are they endangered?" Unfortunately, it is impossible to give a simple "yes" or "no" answer.

First, it is important to understand that the word "endangered" has a technical, biological definition, as do "threatened" and "vulnerable." In this book, we will use the definitions arrived at by the Committee on the Status of Endangered Wildlife in Canada (COSEWIC) (see Appendix B), the body of government and non-government representatives responsible for officially classifying wildlife at risk in Canada. Scientists and wildlife biologists are particularly sensitive to how these words are used. As authors of this book, we chose the subtitle "Predators in Peril" with care. The phrase "in peril" has no technical meaning according to COSEWIC definitions. We chose it because we wanted a phrase to encompass the entire spectrum of conservation concern, one that would allow us to include those wild hunters that are still abundant but won't be for long unless, as the Canadian dean of wildlife biology Ian McTaggart-Cowan says, "we *resolve* that they shall."

Figure 1: PREDATOR STATUS SCALE

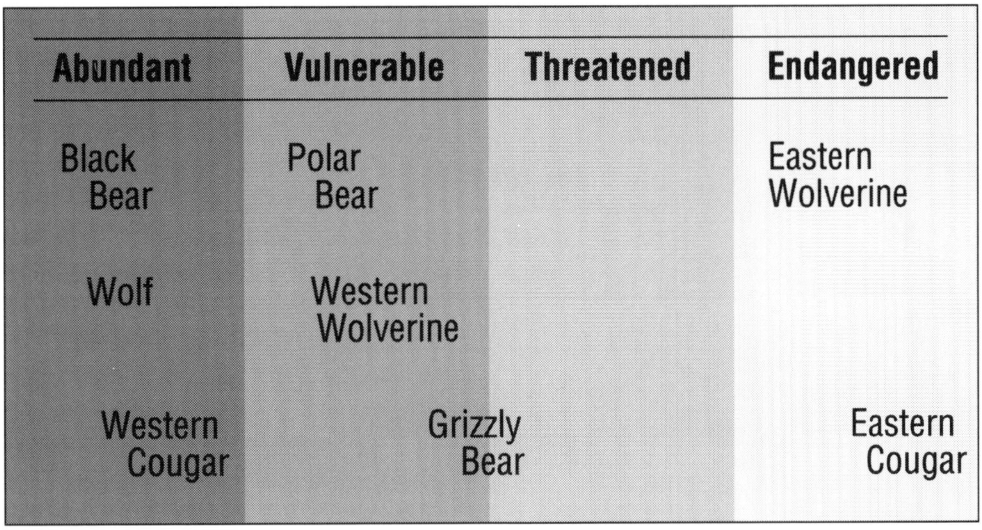

Abundant	Vulnerable	Threatened	Endangered
Black Bear	Polar Bear		Eastern Wolverine
Wolf	Western Wolverine		
Western Cougar		Grizzly Bear	Eastern Cougar

Figure 1 uses a simple "Predator Status Scale" to show that the six large carnivores in Canada are all "in peril" of one kind or another, spread across the status spectrum from abundant, to endangered.

Starting on the left side of the scale with those predators that are still relatively abundant, we find black bears. Also in this part of the spectrum, but not as abundant as black bears, are wolves. Moving farther to the right, we find the western cougar, which is still found in small, healthy populations confined to Alberta and British Columbia.

Polar bears have now been formally designated by COSEWIC as "vulnerable." They are found in low numbers over large areas, but are generally considered secure in Canada. A significant number of them are killed every year by people, however, which means that this species should be monitored carefully. The western wolverine, which lives west of Hudson Bay, is also "vulnerable" by virtue of its natural biology; it, too, is found in low numbers over large areas.

The grizzly bear is a tough one to position on the status scale. We have placed it in transition on the scale between "vulnerable" and "threatened" because it, too, is generally found in low numbers over a large area. In fact, the plains grizzly bear no longer exists in Canada or in the United States, and all other Canadian grizzly sub-populations have been formally classified as "vulnerable." At least one subpopulation has been recommended to COSEWIC to be "upgraded" to "threatened" status, and five others are being killed by humans at a rate which the populations cannot sustain. There-fore, we suspect that some Canadian grizzly bear subpopulations will be formally placed in the "threatened" category in the future. Some experts might argue that a few grizzly bear populations should be in the "abundant" zone of the scale because, even if they're not plentiful (grizzly bears never are), they may be at least as abundant as they've ever been in some places.

Moving farther to the right, we find the eastern wolverine, found east of Hudson Bay. It was formally classified as "endangered" by COSEWIC in 1989, which means the wolverine in this area of Canada could become extinct. And, finally, on the far right, we find the eastern cougar, also classified as "endangered" by COSEWIC. In fact, many experts believe the eastern cougar is "extinct," meaning it may no longer exist at all.

The predator status scale explains why a "yes and no" answer must be given to the question "Are top predators endangered?" Clearly, some are and some aren't, but they're all "in peril" and, therefore, deserving of concern. The scale also shows that the current status of this family of wild animals may be one of the least-appreciated "conservation sleepers" in Canada today. Top predators that are already endangered are obviously a conservation priority because, if we don't tend to their conservation now, they stand to be lost forever. Those that are threatened must also be a conservation priority because, if we don't act with respect to them, they are the next in line to become endangered. Those predators that are vul-nerable are a priority because they can be overexploited and quickly pushed into the "threatened" or "endangered" categories. And those that are abundant or secure for now must become a priority because we still have a chance to do things differently with them, and thereby maintain some of the last wild, viable populations of these magnificent animals to be found anywhere in the world.

Canadian historian Bruce Litteljohn wrote, in *Endangered Spaces*, "Canadians are rapidly approaching the end of the time available for setting aside wild country. If we do not, or if we allow the erosion of existing wilderness places, we will do violence to an important part of our heritage and identity."

Canada is a country that has been defined by the presence of wilderness. It breathes through our history, art, poetry, music, and literature. Our provincial, territorial, and national emblems and our coinage proudly display native wildlife, including some of the top predators that so powerfully symbolize wild Canada. Pierre Elliott Trudeau, in 1944, contemplating the significance of a wilderness canoe trip he had taken, wrote, "I know a man whose school could never teach him patriotism, but who acquired that virtue when he felt in his bones the vastness of his land."

The current status of the polar bear, the grizzly and black bear, the timber wolf, the cougar, and the wolverine indicates how we are doing at harmonizing human activities with nature, and at maintaining our Canadian wilderness heritage. They are the barometers of our success or failure. The purpose of this book is to make sure we succeed.

2.
The POLAR BEAR

To me, the wild polar bear is the Arctic incarnate. When watching one amble

across the pack ice, looking about and periodically sniffing the wind, there is an

overwhelming sense that it belongs there. The Arctic is not a forsaken wasteland

to a polar bear; it is home, and a comfortable home at that. For thousands of

years, the climate, the ice, and the seals upon which it feeds have shaped and

finely tuned the evolution of this predator so exclusively that it has become not

just a symbol, but the very embodiment of life in the Arctic.

IAN STIRLING, FROM HIS BOOK, *POLAR BEARS*

EVERY YEAR, THOUSANDS OF TOURISTS VISIT CHURCHILL, MANITOBA, known as the "Polar Bear Capital of the World." The great bears congregate here every fall, on the shores of Hudson Bay, before freeze-up, and at that time they are easily viewed and photographed. Most of us, however, are able to see live polar bears only in a zoo, swimming in artificial pools or slumped in sleep on concrete pads. Yet, although very few people have the chance actually to see these bears in the wild, virtually all of us like to know that "they're still there."

Inuit residents in the Canadian Arctic still hunt the polar bear, which commands a high place in their culture and economy. Oil drillers, airplane pilots, and ship captains frequently encounter these predators. And wildlife researchers travel over the ice to record bear movements and behavior.

The result of all of these different encounters with polar bears is what bear biologist Charles Jonkel calls "The $500,000 Polar Bear" — a bear treasured by many people for different reasons, people who have each seen "their own bear," and placed their own particular value on it, usually a value far beyond dollars. And all of them, for these different reasons, have an interest in the polar bear's long-term conservation.

POLAR BEAR FACTS AND FIGURES

Those who have had the opportunity to watch a polar bear in its natural environment are often surprised by its size. The polar bear is the largest living land-based carnivore in the world. The hide of an adult male can reach nearly 3.5 meters (11 ft.) in length — equivalent to the length of a compact car. A live bear is an awe-inspiring sight, standing up on its hind legs. The largest males can weigh up to 800 kilograms (1780 lbs.), exceeding the largest male grizzly bears in size. In shape, however, the two species are quite different. Polar bears have a longer neck and legs, and the head and body are more streamlined than those of a grizzly.

The polar bear's shape reflects its adaptation to swimming, and accounts for its Latin or scientific name, *Ursus maritimus*, which means "bear of the sea." The front legs and paws do the paddling, and the hind legs serve as a rudder. In fact, the polar bear is so much a creature of the sea that, in the United States, it is protected by marine-mammal legislation, which also covers whales and seals.

In addition to being sea animals, polar bears are built for living in the cold. Says Ian Stirling, a Canadian Wildlife Service scientist and a world authority on polar bears, "The polar bear is an integral part of the Arctic environment. It's so comfortable and at home, while we're up there just trying to survive. It wouldn't even occur to a polar bear that there would be a better place to be!"

One adaptation that helps these animals survive in the North is their fur. The polar bear's coarse coat normally appears creamy to yellowish white. (Usually the older bears have more of a yellow hue to their coats than do younger individuals.) The fur itself is made up of translucent hairs that allow ultraviolet radiation in. The fur also serves as a "wetsuit," trapping heat while the bear swims in the frigid northern waters. Other adaptations to the cold include thick, insulating hairs around their paws, a dense woolly underfur, small rounded ears to minimize heat loss, and a thick layer of fat under their black skin.

Although much is made of these features that help the polar bears conserve body heat, in summertime it is often more important for the animals to get rid of it. They pant like dogs, losing heat through the black skin of their tongue and lips, and through their ears, footpads, and snout. Overheated polar bears may die of heat prostration, especially if they overexert themselves. Experienced researchers take great care to avoid overstressing these bears while tracking them, especially in summer.

Worldwide, there are probably more than fifteen distinct subpopulations of polar bears. There are twelve in Canada, alone. It is not yet clear how many subpopulations stretch from eastern Greenland east across the Soviet Union to the Bering Sea, but it seems likely it could be anywhere from three to five or more. The worldwide distribution of these subpopulations includes Wrangel Island (USSR) and western Alaska; northern Alaska; the Canadian Arctic; Greenland; Svalbard – Franz Josef Land (Norway); and central Siberia. Our international goal must be to protect all of these polar bear populations.

In Canada, twelve Polar Bear Management Zones have been set up in an attempt to conserve polar bear subpopulations spread across the High Arctic Islands and Hudson and James bays. These subpopulations are thought to be separated by land barriers and open water. Subpopulations also reflect the fact that polar bears are not evenly distributed across their range. Differences in distribution are attributable to variations in the availability of food from one area to the next. Water-filled ice cracks, called "leads," and year-round open-water areas called "polynyas," are rich in seals, the polar bear's primary food. In contrast, in other areas, ice conditions or low food supply cause few seals to be present. Where seals are plentiful, there are likely to be overlapping polar bear home ranges, hence what is called a "subpopulation." As in the rest of the world, in Canada conservation strategies must be designed to maintain not only the species but its various subpopulations.

One concern that must be taken into account when designing a conservation strategy for the polar bear is its relatively low reproductive rate. Females do not reach sexual maturity until four, five, or even six years of age, and, because the cubs stay with the mother for about two and a half years before setting out on their own, adult females breed only every third year, on average. The common litter size is two cubs, although litters of one to four (rare) have also been observed. The survival rate of young, including cubs, yearlings, and subadults, until they first become independent at two years or older, probably varies between about 30 and 50 percent, depending on the size and density of the subpopulations.

Some polar bear populations appear to have a higher reproductive capability than do others. For example, in the lower Hudson Bay region, the average litter size is believed to be higher than elsewhere, and some females appear to breed every two years. Nevertheless, if the average female polar bear lived for twelve to fifteen years, becoming sexually mature at six, and produced two cubs every three years thereafter, she would still produce only four to eight cubs in her entire lifetime, with fewer than that actually surviving to become adults. And, of course, chances are that only half of these would be females capable of producing young. Thus, some female polar bears may not even "replace themselves" unless they are able to live out their normal lifespan.

Wolves, and perhaps black bears, may have larger and more frequent litters as more food and space become available. But polar bears appear to be fairly constant in their reproductive rate, which means they have less ability to increase reproduction under good conditions to compensate for losses in their numbers during difficult times. It appears that the single most important factor for increasing a polar bear population is to maintain and help increase the number of female bears which survive.

Another concern for polar bear conservation efforts is the protection of denning sites. Pregnant polar bears usually den on land, in mid-October to mid-November, in order to provide shelter for their newborn cubs, which are usually born in December. When they become sexually mature, young females will often return to the area where they were born to give birth to their own cubs. Where denning sites are relatively concentrated — for example, on Wrangel Island, Svalbard, and the area southwest of Cape Churchill, Manitoba — they should be protected. In other cases, where dens are more widely distributed over a larger area, such protection may be more difficult to provide.

One fascinating aspect of polar bear reproduction is being researched by Malcolm Ramsay, who has studied polar bears in the Hudson Bay region, with WWF support: "Female polar bears fast for many months during pregnancy and while their bodies prepare to produce milk (lactation). Some of the very largest whales and seals do this, but not for nearly so long, and they certainly don't fast through gestation and lactation." Ramsay wonders how polar bears can fast for eight months without experiencing serious physiological complications, particularly when the fasting and pregnancy occur partially during hibernation. This finding is even more surprising when the way polar bears hibernate is taken into account. Unlike a ground squirrel, which shuts down its body-core temperature very close to zero, the polar bear maintains its normal body temperature during hibernation and is perfectly aware of what's going on around it. If you were near its den, for example, it would know you were there.

Because the polar bear is a hunted species, all aspects of its reproductive biology must be considered when planning conservation measures. Great care must be taken not to kill too many females from the population. Researchers have calculated that, if more than about 1.6 percent of the total number of adult females in a polar bear population are killed annually, the population will decline. Recovery of an overexploited population would then be slow, taking many years. Even unhunted populations may take twenty years to double in size.

The food sources of polar bears are also of interest to conservationists. By far the most important prey species for these bears is the ringed seal, although they will also take bearded seals and occasionally harp and hooded seals as well as walruses, beluga whales, and narwhals. Polar bears catch seals with their massive canine teeth and large, well-clawed forepaws. Their main hunting technique is "still hunting," which involves lying next to a seal breathing hole in the ice, waiting for a seal to surface. Occasionally, they will catch seals by swimming underwater close to the edge of open leads and surfacing at the place where a seal would re-enter the water. They also stalk seals across the ice and break into the tops of seal birthing lairs, or "aglus," to kill the seal pups and any adult females that don't escape. Young polar bears have been observed "practicing" their aglu-smashing technique by rearing up on their hind legs, then coming down hard on the ice or snow with their front legs stiffened. The late Sam Raddi, from Inuvik in the western Arctic, has described a polar bear grabbing a seal by the scruff of the neck, then knocking it back and forth on the ice to kill it before eating.

The most valuable part of the seal to the polar bear is the blubber, because of the high caloric content of its fat and lipids. Sometimes only the blubber is eaten, leaving the rest of the carcass for arctic foxes, which often accompany the bears across the ice for this very reason. Wild polar bears have been recorded as eating an average of 4.4 kilograms (9.7 lbs.) of seal fat and meat per day. In fact, seals are so important to polar bears that much of polar bear conservation work focuses on the ecology of seals. For instance, scientists monitor for the levels of toxic chemicals in seals because such pollutants are transmitted to the bears. The problem is made worse by the fact that these chemicals tend to be concentrated in the fat layers of the seals, the preferred diet of these bears.

In summer, when the ice melts and the bears must come ashore, mainly on Hudson and James bays, they fast until the ice freezes up again in the fall. To a small degree, they feed on sea birds (mostly waterfowl), eggs, small rodents, grasses, sedges, berries, mosses, marine algae, arboreal lichens, and even seaweed. Polar bears will also eat carrion, particularly whale carcasses, and garbage in camps or small settlements. "Garbage bears" used to be a serious hazard in Churchill, but the polar bear-management program there has significantly reduced the problem.

Since these bears are sea mammals, many people think they eat fish, but such is generally not the case. Nevertheless, American bear expert Charles Jonkel mentioned to us that biologists have observed a small population of polar bears that may have been feeding on Atlantic salmon alongside black bears in the Hamilton River/Inlet area on the east coast of Labrador — the only place in the world where this phenomenon has been noted.

There are widespread stories, particularly among the Inuit, of polar bears using a piece of ice as a tool to kill walruses and seals. Jonkel reports seeing a polar bear flip a flat piece of rock end over end for about 6 meters (20 ft.), to trip a bear snare, so it could steal the bait without getting caught. Another researcher, Henk Kiliaan, traced a chunk of ice that had been flipped over and over, to a seal lair. The hole was broken open, but the pup had not been caught. Based on the tracks in the snow, it seemed the polar bear had deliberately rolled the ice piece over and over to the hole. Whether or not the bear picked it up and hammered the hole isn't known, but the chunk of ice was by that hole!

Researchers tell many fascinating and often humorous stories about polar bears. Polar bears don't like to put their feet on small sharp things, a fact used by researchers to catch bears for marking and release. Little sticks are pushed into the snow around a snare to prevent the bear from putting its paw anywhere but right in the trap. One evening, Ian Stirling quietly watched a young polar bear contort its body and lean at an impossible angle in order to avoid sticks placed around the snare-tripping device. The bear stretched out its neck and teetered to the point of almost falling over on its face in order to steal the bait.

Here's another story from Ian: "Cubs stay with their mother for two and a half years, as it takes a long time for polar bears to learn all the things they need to know to survive. Yearlings don't tend to hunt very much, so, often, in the summertime, when the mother is hunting seriously, they're goofing off. This one particular cub was spending a lot of time just taking long runs, doing running leaps and dives into melt pools out on the sea ice. One time it was literally in midair, and I just happened to be watching through a telescope, when a subadult seal popped up in the water and literally implanted itself in the bear's mouth! The poor bear cub didn't know what to do. It was racing around all over the ice, back and forth, throwing this thing up in the air, until its mother finally saw what the cub was doing. She came over and got them back to the business of eating."

These stories make polar bears seem almost human. Perhaps too human. Polar bears can be a problem for people who allow themselves to get into dangerous situations with the bears in the belief that they are "Gentle Bens." However, researchers also object to the image of polar bears as big, dangerous, nasty killers seeking out humans anytime they can. Malcolm Ramsay has handled several hundred polar bears and has never been threatened or attacked. He has concluded that they aren't terribly aggressive toward humans at all. Ian Stirling, who has worked with polar bears for more than twenty years, says, "Polar bears are pretty good natured. I liken working on them to driving in traffic. You have a real responsibility to make sure that you don't put yourself and the bear in a situation where it might be really dangerous. As long as you pay attention to the red lights and the green lights, so to speak, you won't have much in the way of problems."

Malcolm and Ian's experiences indicate the importance of respecting polar bears, and taking basic safety precautions when they are around. One resident Hudson Bay goose hunter told us he nearly stepped on a sleeping polar bear while putting out decoys in the dark one morning. The white bear looked like a patch of snow in the dim light. Similarly, anyone roaming the Hudson Bay willow flats; bird watching; visitors hiking the beach ridges of Polar Bear Provincial Park in Ontario; Inuit seal hunters on the open ice — all need to keep an eye out for, and know how to avoid problems with, polar bears. The Inuit traditionally used dogs to chase polar bears when they were hunting them. Today, one of the best alarm systems for roaming polar bears is still a dog or two tied up near the edge of camp.

Polar bears are curious. They often investigate a strange object in open country by first approaching it, then examining it by smelling it, holding it, and perhaps even chewing on it. As a result of this behavior, "attacks" have occurred when people have tried to feed bears by hand at dump sites or from wildlife-observation vehicles. In one fatal attack in Churchill, it was later revealed that the man had carried food in the pockets of his coat. A little common sense can go a long way to insuring human safety, as well as that of the polar bear.

WHAT HAS HAPPENED TO POLAR BEARS?

The worldwide range of the polar bear includes territory under the jurisdiction of five different nations: Greenland, Norway (Svalbard only), the Soviet Union, Canada, and the United States. These animals are also found in the ice-covered waters of the Beaufort, Chukchi, East Siberian, Laptev, Kara, Barents, and Greenland seas, plus James, Hudson, and Baffin bays, and Davis, Hudson, and Bering straits.

In Canada, polar bears are most abundant in coastal areas where annual ice with leads and polynyas occur — Beaufort Sea, Amundsen Gulf, Hudson Bay, Foxe Basin, Baffin Bay, and the Eastern High Arctic. Bear densities are lower in areas such as the Coronation Gulf where smooth, continuously frozen ice occurs annually. And the fewest bears are found where multiyear ice predominates, such as in Viscount Melville Sound and around the Queen Elizabeth Islands. On balance, however, and unlike other bear species, the polar bear still inhabits more or less all of its original habitat (see Figure 2).

There is a major misunderstanding about the status of polar bears. Many people mistakenly believe they are an endangered species. Ian Stirling explains how this misconception came about: "It was legitimately thought that polar bears might be becoming endangered in the early 1960s, because harvesting became extremely efficient with the introduction of the snowmobile to the North. This, along with high-powered rifles, rapidly rising prices (for polar bear hides), and no quotas, simply set the stage for very rapid increases in harvests in Canada and elsewhere. There was also relatively unrestricted hunting of polar bears by aircraft in Alaska. In the vicinity of Svalbard, they were shooting swimming polar bears and trapping large numbers, and so on. So, if you looked at the worldwide numbers that were being recorded as being killed (you would have no way of correcting for the unrecorded kills), it was pretty clear that things were not looking good."

As examples of what Stirling is referring to, trophy kills in Alaska increased from 139 in 1961 to 399 in 1966; recorded kills in Canada went from 350 to 550 from 1953 to 1964, and jumped to 726 kills in 1967. Set gun traps were being used in Svalbard, Norway. The "hunter" would attach bait to the trigger of a firearm, which then

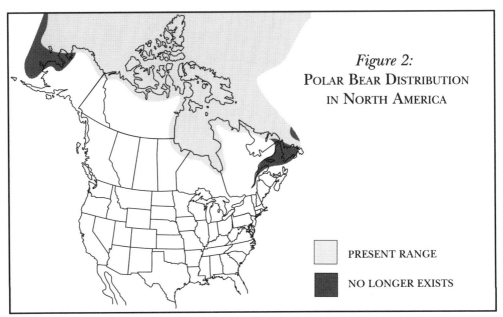

Figure 2:
POLAR BEAR DISTRIBUTION
IN NORTH AMERICA

PRESENT RANGE

NO LONGER EXISTS

Source: M. Novak, J.A. Baker, M.E. Obbard and B. Malloch, © 1987, Queen's Printer for
Ontario (Toronto: Ontario Ministry of Natural Resources and Ontario Trappers
Association), p. 476.

indiscriminately wounded or killed any bear that set it off, male or
female, cub or adult. As a result of such practices, some experts during the 1960s were estimating the world population of polar bears to
be as low as 8,000 animals. No wonder the alarm bells were ringing!

In 1965, meetings of representatives from all five nations with
polar bears were held in Alaska, and agreement was reached on a
number of preliminary points relating to conservation. This
agreement, in turn, led to the formation of the International
Union for the Conservation of Nature and Natural Resources
(IUCN) Polar Bear Specialist Group in 1967. It has met more or
less every two years since 1968, and in 1973 negotiated the International Agreement on the Conservation of Polar Bears and
Their Habitat (see Appendix C).

The IUCN agreement is not strictly protectionist, which probably accounts, in large part, for its success. Rather, it sets conser-
vation-sensitive conditions under which polar bears can be killed.
The polar bear agreement also called for improved national
research programs in each country for the purpose of conservation. But, most important, the signatory countries agreed to

protect the Arctic ecosystem, of which polar bears are a part.

Additional specific conservation measures related to polar bears are worth noting. For example, there has been a complete ban on hunting polar bears in the Soviet Union since 1956, and, in 1976, Wrangel and Herald islands were designated as state reserves by the USSR.

Hunting in Greenland is now confined to native Greenlanders. All cubs up to one year old are protected, as are denning areas up to 20 kilometers (12 mi.) out to sea. In 1973, Northeast Greenland National Park was established.

In 1965, it became illegal to shoot polar bear cubs and females in Svalbard; in 1967, hunting from snowmobiles, boats, or aircraft was prohibited; and, in 1973, the Norwegian government established a five-year moratorium on the hunting of all polar bears in Svalbard — a moratorium that has never been lifted. Forty percent of Svalbard was protected in 1973 through national parks, nature reserves, and bird sanctuaries.

In 1972, the United States passed the Marine Mammal Protection Act, which banned all hunting of polar bears in Alaska, except by aboriginal people for subsistence purposes.

As a result of these measures, the polar bear populations of the Soviet Union and Greenland are considered to be secure; the population is thought to have doubled in Svalbard between 1973 and 1983; and the Alaskan population has probably increased slowly since guided hunting of polar bears was stopped in 1972.

Canada has participated fully in the IUCN Polar Bear Specialist Group and is a full signatory to the International Agreement on Polar Bear Conservation. In 1968, the Northwest Territories introduced polar bear-hunting quotas to all its arctic villages, and committed to do further research to determine whether the quotas were too low or too high. That led to the formation of the Polar Bear Technical Committee, made up of representatives from the four provinces and two territories with polar bear populations, plus the federal government, to conduct conservation research in Canada. A Polar Bear Administrative Committee, made up of the wildlife directors from the provinces, territories, and the federal government, was also formed to make management decisions based on that research.

Today in Canada, only Inuit and Indians are legally allowed to hunt polar bears, and they hunt within an overall quota system. In the Northwest Territories, a certain number of hunting tags are distributed to each arctic community. It is left up to the local Hunters and Trappers Association (HTA) to determine distribution of the tags to individual hunters within the community. Native hunters may allow some of their tags to be used by a non-native person on an Inuit-guided polar bear hunt. The price for polar bear hides is $1,000 to $2,500, whereas these guided hunts are worth about $18,000 to $20,000 each to the resident with a polar bear tag. Although there is a strong cash incentive to use the tags in this way, the honor and tradition associated with killing a polar bear are still strong enough in many Inuit communities to limit this non-native use of the system. From a conservation viewpoint, the same number of bears are killed, within the overall quota, whether they are taken by native or non-native hunters. At HTA meetings in the Keewatin District, a decision was made to count wounded or "lost" bears in the quota. Also, the HTAs of Broughton Island and Clyde River voted to reduce their quotas by two-thirds for seven years when population studies indicated that the polar bears there were being overhunted. These are examples of responsible, locally initiated conservation moves.

In 1970, Polar Bear Provincial Park was established in Ontario — the second-largest park in Canada and one of the largest in the world. This provincial park includes polar bear denning sites. In 1968, Manitoba established wildlife management areas at Cape Tatnam and Cape Churchill. Cape Churchill is now a candidate for national-park status as well. Additional protected areas in the form of national parks and ecological reserves have been proposed for the Canadian Arctic, which would further protect polar bear habitat.

As a result of these conservation measures, polar bear subpopulations in most of the Canadian Management Zones are believed by scientists to be stable. However, the status of those in Foxe Basin, northern Hudson Bay, and northern Quebec and eastern Baffin Island is uncertain. It is possible that quotas may need to be readjusted after new population studies have been completed. In the 1970s, there was concern about declines in the Beaufort Sea and western Arctic, but this population now appears to be secure.

How many polar bears are there? This question is greeted with frustration by polar bear experts for two reasons. First, they don't know, and, second, they're not sure it's important — the key for practical conservation purposes is not to guess at a world population number, but to home in on the status of subpopulations. On this question, Ian Stirling says, "In a sense, it's not really much more than a *Guinness Book of Records* kind of fact to know how many there are. It's how many there are in each area in which they're either being hunted or being impacted upon by some other activity that's important."

Table 1 gives Stirling's estimates for various subpopulations of polar bears in the world, taken directly from his 1988 book, which summarized information available at that time. The total world population from this count would be 19,175 polar bears. However, these are estimates for only those subpopulations where surveys have been undertaken. Some of the surveys were not recent, and data for several areas were not adequate to allow for an accurate estimate. More to the point, these estimates exclude polar bears from the Foxe Basin, the east coast of Baffin Island, and the northwestern High Arctic in Canada, as well as bears from the Chukchi Sea in the Soviet Union, where estimates range from

Table 1:

ESTIMATES OF THE SIZE OF SUBPOPULATIONS OF POLAR BEARS WORLDWIDE

Area	Estimate
Barents and Greenland seas	5,000
East Greenland	200
Baffin Bay and Thule district	300
Canadian Arctic	
Zones A1 and A2 (southwestern Hudson Bay and James Bay)	1,500
Zone A3 (western Hudson Bay)	1,500
Zone B (Labrador coast)	75
Zone D (south)	700
Zone E (Central Arctic)	1,100
Zone F (High Arctic)	2,000
Zone H (eastern Beaufort Sea, Amundsen Gulf)	1,200
Alaska North Slope and western Zone H	2,000
Soviet Union	3,600
Total	**19,175**

Source: Ian Stirling, *Polar Bears* (Ann Arbor: The University of Michigan Press, 1988), p. 77.

2,500 to 7,000. Stirling suggests that the polar bear estimates in Table 1 from the Soviet Union are likely too low. If proper counts were done for the Soviet Union, and if other unsurveyed areas were included, he says, "It seems quite possible that the world population of polar bears could be approaching 40,000." This figure may be overly optimistic. The IUCN Polar Bear Specialist Group in 1981 estimated the world population of polar bears to be between 20,000 and 40,000. And Chris Servheen, co-chairman of the IUCN Bear Specialist Group, cites sources that would put the worldwide number at approximately 25,000 bears. Take your pick!

In any case, Canada likely has about 50 percent of the world's population, and therefore may harbor somewhere between 12,500 and 20,000 polar bears. In the 1991 COSEWIC status report, the Polar Bear Technical Committee estimates Canadian polar bear numbers to be on the lower side of this range, at approximately 13,000 to 15,000. This gives Canada a great responsibility for conserving this magnificent animal, which has captured the world's imagination as a powerful symbol of the Arctic.

CURRENT THREATS TO POLAR BEARS

It is important both to put the good news regarding polar bears into a more sober context and to measure the bad news against a backdrop of some real conservation success stories. In other words, we shouldn't be complacent, but neither should we exaggerate concerns. The conservation status of polar bears is confusing to many people because some experts emphasize the positive side of the picture, and others stress the negative side. Such confusion has been further reflected in Canadian attempts officially to classify the polar bear, which has defied standard designations such as "endangered," "threatened," or "vulnerable." With this in mind, and in an effort to be factual, here are some of the current threats to these bears.

Throughout the Canadian Arctic, about 700 polar bears are killed every year by Inuit and Indian hunters. This figure could be exceeding 5 percent of the total Canadian polar bear population, and is very high for such a slow-reproducing animal. The maximum level of hunting kill for grizzly bears, for example, is judged to be 3 percent. When calculating percentages, however, it is more important to consider numbers of bears taken out of each subpopulation to determine whether the polar bear in Canada is being overexploited. Most important is the proportion of adult females within each subpopulation.

The problem is that we don't have reliable estimates of the size of most subpopulations. Without such estimates, it can't be definitively stated that hunting levels are safe or unsafe in all areas. For example, about 200 polar bears are taken each year from the area of northern Hudson Bay, Foxe Basin, Hudson Strait, and the Labrador Sea. These bears may come from more than one subpopulation because the boundaries are not known, and they are hunted by Inuit from four different user groups. Currently there is not enough known about the size of this subpopulation of polar bears, or its boundaries, to understand the cumulative effect of this level of hunting. In Quebec, under the James Bay Agreement, there are guaranteed levels of harvest for polar bears, but these are not based on biological information regarding polar bears. Research will continue, and there are plans to co-ordinate the harvest levels of all the user groups, but as you can see, it's complicated!

The short-term response to establishing hunting quotas for polar bears in the Northwest Territories has been to base them on historic levels of harvest. That usually means allowing hunters to take the same number of bears they have in the past, on the understanding that these levels will be adjusted up or down, depending on the findings of new research and fieldwork. While there may not have been changes in the numbers of bears taken, there have been significant changes in hunting practices. For example, females and females with cubs have been protected by setting hunting seasons at times when females are less likely to be killed and by protecting all bears in dens.

In some cases, polar bears of the same subpopulation roam back and forth between government jurisdictions that may be managing them quite differently. The polar bear subpopulation in the southern Beaufort Sea, for instance, moves between Canada and Alaska. On the Canadian side, strictly enforced quotas for Inuit and Inuvialuit settlements were established in 1968. In Alaska, it was legal for the Inupiat people to kill bears for subsistence purposes with no limits, and there was no protection for females with cubs or for bears in dens. In other words, for the same subpopulation of polar bears, killing was controlled in part of its range but not in another. Furthermore, under the U.S. Marine Mammal Protection Act, the government still can't take effective conservation action until the Alaskan polar bear population is shown to be depleted! In this case, when governments seemed unable to act cooperatively in the interests of conservation, the native people themselves took the initiative. The Inupiat of Alaska and the Inuvialuit of Canada signed their own conservation agreement in 1988. If this agreement works, it could be an important model for conservation efforts by those who have the greatest interest in maintaining shared polar bear populations.

Lucrative non-native polar bear hunts could give rise to a demand for a larger quota, resulting in a potentially negative conservation impact on polar bears. In some cases, this has already occurred. In other cases, this form of hunting has had a very different effect because not all sport hunts result in the killing of polar bears. If the hunt is unsuccessful, the tag allotted to that hunt cannot be returned to the village quota system and used again. As a result, fewer bears are being killed than the quota system would normally permit.

Although there have been concerted efforts by some governments and by aboriginal peoples to keep hunting quotas within safe conservation levels, pressure comes from the communities to increase the quotas or to distribute them differently among settlements. The latter is important because it could shift hunting pressure from one subpopulation of polar bears to another. Finally, the figure of 700 bears killed annually is close enough to the upper permissable level that any additional deaths — from an environmental disaster such as an oil spill or from killing too many problem bears that wander into camps — could push some subpopulations into decline. Recovery from such a population decline could take a long time, given the fact that polar bears are slow in reproducing, assuming that the pressures causing the decline in the first place could be eased.

Some killing of polar bears for the defense of humans and property is unavoidable. The key is to reduce the number of bears killed for these reasons. There are precautions people can take to avoid problems with polar bears, and Canadian governments do an excellent job of advising Arctic visitors and residents about what they can do in this regard. In some cases, bears can be deterred by non-lethal means, such as noisemakers and harmless rubber bullets, or by relocating them after they have been tranquilized. Nuisance bears that must be killed should be included in the overall quota and thereby taken into account when determining the total number of bears that may be killed through hunting. Most of the time, problem-bear kills are included in the quota, but the number is difficult to determine in advance because bears are killed in self-defense in variable and unpredictable numbers. As long as this problem persists, we run the risk of inadvertently killing too many bears and possibly causing population declines.

Industrial activities pose another threat to polar bears, particularly if combined with hunting and killing of problem bears. Since the 1970s, there have been sporadic but repeated efforts to explore for, and produce, oil and gas from the Arctic Continental Shelf. Threats to polar bears associated with these activities include: oil spills from well blowouts, tanker accidents, or pipeline ruptures; pollution from various by-products associated with drilling; noise from seismic operations and ice-breaking; and changes in ice conditions caused by boat traffic. Direct contact with oil slicks affects the ability of polar bears to survive the cold and would likely lead to death in an arctic environment. Polar bears often lick their paws and wash their faces in a pool of water after feeding on seals.

If the water is contaminated by oil, or if the bears eat oiled seals or carrion, they could poison themselves. One wild polar bear at Churchill recently drank about a gallon of hydraulic fluid from an open pail before being chased away. The fate of the bear is unknown, but the incident indicates that polar bears will not always avoid oil.

All top predators are particularly susceptible to toxic chemicals in the food chain through a process called "bioaccumulation." In other words, as you move up the food chain, animals at each level accumulate higher levels of toxic chemicals in their bodies than those animals below them. Obviously, through bioaccumulation, those animals at the top of the food pyramid, particularly summit predators such as polar bears, amass the highest levels of toxic chemicals.

Although contaminants may originate thousands of kilometers away, they are being found in the oceans and atmosphere of our entire planet, including the Arctic. Mercury has been discovered at elevated levels in polar bear hair and in seals in some parts of the Far North. It's not clear, however, whether the source of mercury is natural or man-made, or whether these elevated levels are actually a hazard to the bears. We do know mercury is a hazard to humans. High concentrations of polychlorinated biphenyls (PCBs) — a controversial chemical used extensively in electrical transformers — and DDT — a pesticide restricted in use during the 1970s — have also been found in polar bears. PCB contamination appears to be a particular threat to cubs because of high concentrations of the chemical in the milk of the mother bears. PCB levels drop through the middle years of a polar bear's life, and then elevate again as it gets older as a result of eating many contaminated seals over its lifespan.

Measurements comparing toxic-chemical levels from the late 1960s to the 1980s indicate an increase. Levels of the pesticide chlordane, for example, doubled in polar bears in Hudson and Baffin bays over that period of time. These findings are causing concern about the overall health of polar bears, and of residents in the Arctic, who are also at the top of the food chain.

When asked what he thought were the most serious threats to polar bears, Malcolm Ramsay said, "Probably the biggest threat is the possibility of global warming caused by massive emissions of 'greenhouse gases,' particularly carbon dioxide, from industrial sources, power stations, automobiles, and the burning of the earth's forests. The models climatologists are putting forward suggest that the greatest warming is going to take place in high latitudes, that is, both the Arctic and Antarctic. If we talk about a one- or two-degree net warming of the earth, that might translate into three or four degrees in the Arctic. A three- or four-degree temperature rise in the Arctic would be catastrophic. It would affect the distribution and abundance of polar bears massively, because it would greatly affect sea-ice conditions. If ice dynamics change, then the whole feeding ecology of the bears could change in ways that are unpredictable. But almost certainly, it would be deleterious to the bears." As is shown by the effects of toxic chemicals and global warming, polar bears are being swept up in a broader environmental problem.

It may come as a surprise that we should think of tourist businesses, built on simply watching polar bears and appreciating them from a distance, as a threat to these great white bears. Such viewing has become a multimillion-dollar industry for the town of Churchill, Manitoba, where there is a concentration of polar bears in the late summer and fall. At that time, ice on Hudson Bay has melted, and the bears come ashore, waiting for freeze-up, when they can move back out onto the ice to catch seals. Churchill thus finds itself located in a remarkable natural congregating place for polar bears. On balance, this has been a good thing for both the town and the bears. Tourists bring in much-needed revenue and leave with a greater appreciation of polar bears. However, it has also led to problems.

Polar bears are seen as posing a potential threat, as more bears and people visit Churchill. Consequently, there is a twenty-four-hour safety patrol during the bear season to protect the town and to prevent bears from being killed when they are perceived as dangerous. Churchill has also built a holding facility ("bear jail") to detain problem bears safely until freeze-up, when they can be released. Another problem is the tourists themselves. They sometimes approach the bears too closely, especially in the area of the town garbage facility, or behave improperly during special tours out on the tundra, for example, by feeding bears from observation vehicles. There is also the possibility of inadvertently harassing bears in a concentrated denning area through frequent viewing visits by people in all-terrain vehicles. The answer to virtually all of these threats is not so much to manage bears as it is to manage people.

BLUEPRINT FOR SURVIVAL

In addition to the conservation measures outlined in Chapter 8, which apply to all large carnivores in Canada, specific steps must be taken to conserve polar bears in particular.

1. *International Cooperation*

Canada should continue to play a constructive and leading role in the IUCN Polar Bear Specialist Group and in the 1973 International Agreement on the Conservation of Polar Bears and Their Habitat. Such participation will insure that Canadian conservation efforts take place in the context of worldwide priorities for this animal. The conditions of the agreement must continue to prevent overhunting and reduce the detrimental effects of humans on the bears' environment, while insuring that adequate research is carried out for conservation purposes.

All five participating nations reconfirmed their commitment to the Polar Bear Agreement in 1981, but compliance to date has been based on the honor system. Legally speaking, the terms of the agreement are not enforceable by any country, and structures for implementing it are up to the individual countries. Simon Lyster, Treaties Officer for WWF, warns, "The fact that parties are not required to hold regular meetings to recommend ways of making the agreement more effective has not yet been a serious hindrance, but it may make it easier to ignore the provisions of the agreement if they prove to be a serious stumbling block to future industrial development in the Arctic."

2. *Canadian Measures*

In Canada, the Polar Bear Technical Committee, with representation from all jurisdictions with polar bear populations, should continue to coordinate conservation research on this species, and the Polar Bear Administration Committee should continue to decide on management in a national context.

3. *Aboriginal Hunting*

Inuit hunters must be more involved in the conservation and management of polar bears. This involvement can be achieved through continued local allocation of quotas by Hunters and

Trappers Associations, through native agreements (such as the one mentioned earlier between the Inupiat of Alaska and the Inuvialuit of Canada), and through further partnership with wildlife officials to involve Arctic residents in polar bear conservation by seeking their guidance on research and fieldwork.

In Alaska, it is still legal for aboriginal people to hunt polar bears for subsistence purposes with no limit on the overall number of bears taken. This situation is of conservation concern because it could lead to overexploitation of a polar bear subpopulation that also happens to be shared with Canada. One hopes that, in the absence of an international agreement between the two countries, the agreement between the Inupiat of the United States and the Inuvialuit of Canada will prevent such overexploitation.

In Quebec, although there is no formal quota system, Inuit hunters have generally accepted guidelines on harvest levels for polar bears. However, there is still conservation concern about the status of this subpopulation of bears, which appears to be shared by northern Quebec, the Northwest Territories, and Labrador. Resolving the problem will require coordination among these different jurisdictions, and there is an urgent need to pull all the users together to negotiate an agreement, perhaps similar to that established by the Inupiat and Inuvialuit.

Female polar bears with cubs should continue to be protected, and all possible precautions should be taken, including adjusting hunting seasons, to protect females. Hunted polar bear subpopulations should be monitored annually to insure that the annual kill remains within safe conservation limits.

In many cases, native hunters have agreed to limit their take of arctic wildlife to historic levels on the condition that further research will be done to determine a level of hunting that is safe. When a commitment of this kind is made, particularly by government agencies responsible for doing such research, it must be honored out of human decency, and because people's livelihoods as well as the future of the bears are at stake. It is not good enough to solve an immediate conservation concern by obtaining such an agreement from village residents, then not delivering on the commitment to allocate funds and expertise to do the promised research. That has happened with respect to many species hunted by Arctic peoples. Failure to honor such agreements makes native hunters understandably reluctant to consider them seriously in future situations, which jeopardizes a major conservation instrument in the North. In general, realistic, well-intentioned, and, in some cases, voluntary steps are being taken to keep the number of polar bears killed by native hunters within safe conservation levels. These efforts should be supported by southern-based conservation groups who may prefer to take a more protectionist stance. Instead of advocating a ban on the hunting of polar bears, a more tolerant approach would be to work together to insure that the number of bears killed is sustainable. That is, after all, in the best interests of both the polar bears and the Arctic residents, who have coexisted successfully for centuries.

4. *Protecting Critical Habitats*
Polar bears move over great distances during the course of a year, because large-scale seasonal changes in the distribution of different kinds of sea ice cause a significant variation in the distribution and abundance of seals. Although polar bears have immense home ranges, it doesn't follow that it is impossible to protect specific areas to help conserve them. High-density denning areas, such as those at Churchill, and on Wrangel Island, Svalbard, and the seaward tips of Cumberland, Hall, and Meta Incognita peninsulas on Baffin Island, must be protected from human interference, as must summer retreats and key feeding areas, especially those around recurring leads and polynyas.

5. *Reducing and Preventing Industrial Impacts*
Proposals for industrial development in the Arctic must be accompanied by adequate environmental-impact statements, which assess the direct effects on polar bears and their habitats. Furthermore, industrial-liability agreements for such impacts, including compensation paid to Inuit hunters for bears lost and for restoration of polar bear habitat, should be spelled out as a condition of such activities. Negotiations in this regard for Beaufort Sea industrial activities could and should set a useful precedent. Where areas of interest to oil and gas companies overlap with critical habitats for polar bears, environmental-impact

studies must be used to determine not just how such industrial development should proceed, but whether it should proceed at all.

Emergency plans to contain oil spills in the Arctic must be in place, particularly when there are potential impacts on areas such as shore leads parallel to the coastline and polynyas frequented seasonally by polar bears for feeding and travel. Government agencies must insure that any polar bear losses as a result of industrial activities are taken into account and included when determining the number of bears that can be killed annually through the hunting-quota system.

6. *Toxic Chemicals and Greenhouse Gases*

Polar bear tissue should be regularly monitored for the presence of toxic chemicals, such as mercury, DDT, and PCBs. The focus of such research should be on tracking the biochemical health of the bears themselves, and on any deterioration of the arctic marine ecosystem of which they are a part.

Controlling the sources of these chemicals involves determining acceptable air-pollution emission levels and pesticide practices elsewhere in Canada and the world, as well as negotiating international agreements regarding the long-range transport of such chemicals. International agreements in this respect are especially important between Canada and the Soviet Union, where industrial development in the North is believed to be responsible for increasing Arctic pollution.

With respect to greenhouse gases and global warming, so far Canada has agreed only to stabilize its emissions of carbon dioxide at 1990 levels by the year 2000. That is not good enough. A more demanding proposal, consistently supported by groups such as WWF and agreed to by many other nations, would require a 20 percent reduction in carbon dioxide emissions below 1990 levels by the year 2005.

It may seem strange to be advocating such measures as the control of agricultural pesticides and air pollution in a chapter on conserving polar bears, but it's a fact that including them here simply reflects the fundamental interconnectedness of our global life-support system. The bulk of the world's people who do not live in our planet's polar regions must become aware of how our consumer preferences and excesses, our daily behavior, and our wasteful resource-disposal practices affect species, ecosystems, and people many thousands of kilometers away.

7. *Research Needs for Conservation*
There are five areas of research that need to be further developed and funded. One, we must better understand the size and boundaries of natural, national, and international subpopulations of polar bears. To do so involves developing more accurate censusing techniques and being committed to doing such work once the techniques are developed. Two, we must continue to keep careful records of the sex and age of all polar bears killed through hunting and problem-bear removals to make sure we are not jeopardizing the reproductive capability of a particular polar bear subpopulation. Three, we must better understand industrial impacts, particularly on critical habitats such as denning and feeding areas, and the effects of crude oil. More importantly, we must take action to make sure negative impacts are avoided. Four, we must further investigate the presence and long-term effects of toxic chemicals on polar bears, and relate these to anti-pollution measures being proposed elsewhere in the world. And, finally, we should develop more effective non-lethal deterrents, and make sure they are used, to avoid having to kill problem polar bears while protecting work camps and tourists.

8. COSEWIC *Designation*
The Federal/Provincial/Territorial Polar Bear Technical Committee prepared the status report on the polar bear for the Committee on the Status of Endangered Wildlife in Canada (COSEWIC). Their recommendation that this species be officially classified as "vulnerable" in Canada was formally approved in April 1991.

It is very important that we all understand what such a designation really means. It does not mean that the polar bear is an "endangered species," "on the brink of extinction"; its numbers are not low enough for that classification. Nor does it mean that the polar bear is a "threatened species," "on the brink of becoming endangered," because, overall, polar bear numbers appear to be stable.

The polar bear has been classified as "vulnerable" for the following reasons: it is found in low numbers over a large area; it has a slow reproductive rate; the size of a subpopulation could decline as a result of overhunting or an uncontrollable environmental disaster; such declines might be difficult to detect until they were in progress and serious; it could take years for a polar bear population to recover from such a decline; we lack sufficient information on subpopulations; and toxic chemicals and global warming have unknown but potentially negative long-term effects on this species.

For much the same reasons, polar bears have already been classified as "vulnerable" on a worldwide basis by the International Union for the Conservation of Nature and Natural Resources (IUCN), and listed on Appendix II of the Convention on International Trade in Endangered Species of Wild Flora and Fauna (CITES). (See Appendix D for an explanation of CITES and the protection afforded to species through the CITES Appendix system.)

Polar bears are beautiful animals, and are of great interest to a broad spectrum of people around the world. In the long run, if polar bears are to be successfully conserved, it will be necessary to fashion agreements among diverse human interests, focusing on subpopulations of the bears. Ian Stirling advises, "The only way we're going to get at this, for each different subpopulation of bears, is for everybody involved with polar bears — hunters, government managers, scientists, tourists, and non-consumptive users — to look at things like population size, sustainable yields, critical habitat areas, harvesting seasons, places where people can go see them, and areas where polar bears should not be disturbed but left to be bears without being bothered by anybody. All those kinds of things need to be brought together into agreements, which are going to be a little bit different in just about every area because the sweep or combination of users and ecological circumstances is going to vary."

Overall, the world's population of polar bears could be described as secure for now, but nothing to be complacent about. Thor Larsen, a Norwegian member of the IUCN Polar Bear Specialist Group, has said, "We must not rest on our laurels with respect to this species." His warning should be heeded. The polar bear is a vulnerable, slow-reproducing top predator. Large numbers are removed every year from the Canadian population through controlled killing by people. For these reasons, this species demands vigilance.

3.

The GRIZZLY BEAR

We saw one older male and several families of females with little cubs. They

were walking along, paying attention to one another, just doing what bears do.

It was very, very beautiful to see them, just to get a glimpse of the quietness of

their usual lives. I felt extremely at ease, part of what was there. It was my first

impression of what the Earth is supposed to be like.

CANDACE SAVAGE, AUTHOR, DESCRIBING A FLIGHT INTO COPPERMINE RIVER
BACKCOUNTRY, NORTHWEST TERRITORIES

IT'S A SAFE BET THAT NOT MANY PEOPLE THINK OF THEIR CHILDHOOD teddy bear as a grizzly bear. Just the sound of the word "grizzly" conjures up images of ferociousness and attacks on innocent people. Yet, these popular impressions, the cuddly teddy versus the rampant killer, are derived from the same species of animal — the great brown bear that once roamed half of North America and nearly all of Europe and Asia.

Today, the grizzly bear is becoming better known for what it really is, an indicator of the human species' willingness or unwillingness to abide by and within wilderness. We are fast determining whether or not there is going to be room for both grizzly and humans on this planet. For the real grizzly, as opposed to the storybook version, the prospects do not look good.

GRIZZLY BEAR FACTS AND FIGURES

Grizzly bears weren't always in trouble. In fact, brown bears (*Ursus arctos*) were the most widespread of any bear species in the world. In North America, there are two subspecies: the Alaskan brown bears or Kodiak bears (*Ursus arctos middendorffi*), which are found in the Alaskan Islands, and grizzly bears (*Ursus arctos horribilis*), which are found throughout the rest of brown bear range in Canada and the United States.

In 1917, American naturalist C.H. Merriam thought there were up to eighty-seven different kinds of grizzlies and brown bears in North America, but these are now recognized as simply variations of one species. If one were to divide grizzlies into groups, it would probably make sense to do it according to the major habitats where they live. Those found in mountainous areas, such as the Rocky Mountain national parks, could be called "montane" grizzlies. Those found feeding in salmon estuaries along coastal British Columbia could be called "coastal" grizzlies. And those inhabiting the largely treeless areas of the northern Yukon and Northwest Territories could be called "tundra" grizzlies. The "plains" grizzly bear, which once roamed the prairies, feeding on bison along with the buffalo wolf, is now extinct. As with all top predators, it is important to conserve not only the grizzly bear, but also these different kinds of grizzlies, and even subgroups within these broad habitats where they still exist.

Peter Clarkson, a wildlife biologist with the Government of the

Northwest Territories, points out that some of these differences may not be entirely distinct. In the case of the barren-ground or tundra grizzly, for example, some of the bears live on the tundra for most of their lives, some live in the mountains, and others go back and forth between the two. In other words, there may not be a distinct tundra population of bears that live only on the barren ground.

Nonetheless, different habitats do account for differences in the size and appearance of grizzly bears. Generally speaking, in the richer environments, where high-quality food is in good supply, the bears are larger. Adult weights thus vary from 68 to 536 kilograms (150 to 1,190 lbs.) across their range. It would take at least six strong men to lift a large sleeping male grizzly bear up off the ground, if they were crazy enough to try!

Grizzly bears are smaller on average than polar bears, and larger on average than black bears, with a more dish-shaped face, longer front claws, and a much more defined shoulder hump. As the name implies, grizzly bears often have a "grizzled" coat, which ranges in color from creamy yellow to honey, through all shades of brown, to nearly black. Like any animal with a long fur coat, when grizzlies are wet, their appearance changes quite dramatically — they look thinner, less massive, and, when standing on their hind legs, amazingly human. Fred Bunnell, professor of wildlife and forestry at the University of British Columbia, notes that, "when they stand up on their hind legs to try and sense their environment better, they look pretty impressive! Some people may see that as a threat, but I think they're just trying to figure out what the hell's going on!"

Individual grizzly bears also vary greatly in their general shape. Some are long and rangy; others are chunky and heavily built; some have slender heads; others have broader, more concave faces. People who have a stereotyped "Gentle Ben" preconception about what a grizzly "should" look like are often surprised by how different they can appear in the wild.

Another stereotype regarding wild grizzlies — and, for that matter, all bears — is that they lack personality because they are thought to spend their whole lives simply surviving. Consequently, bears have been likened to pigs of the bush — males are often called "boars" and females, "sows." However, grizzlies show all kinds of interesting behavior. Many of the bear biologists we interviewed reported watching grizzly bears playfully sliding down snowbanks, doing headstands, and digging holes. Some bears have learned how to steal bait from snares, even when the people who set the snares couldn't place the bait in them without catching themselves! And grizzlies have been reported to be able to identify certain vehicles, even by the sound of the doors slamming, as belonging to people they know. Bear researchers also emphasize the differences between individual bears. Like humans, some bears are more belligerent, or intelligent, or shy, or aggressive, or curious, than others.

Contrary to another popular misconception, grizzly bears, and all bears, *can* run downhill. Stephen Herrero, from the University of Calgary, a world authority on grizzlies and co-chairman of the IUCN Bear Specialist Group, thinks he knows where that one came from: "from the fact that bears appear to have shorter front legs and appear to be slope-backed. Therefore, someone down the line must have thought that bears would stumble running downhill. It's completely untrue; I have seen bears run downhill full-tilt, and turn on a dime!"

Adult grizzlies are usually solitary, except during the mating season, though small congregations may be found around particularly rich seasonal feeding areas such as salmon runs or dense berry patches. These feeding spots are often well known to both grizzlies and grizzly bear watchers. Such areas are important for conservation purposes because they are examples of critical grizzly bear habitats that must be protected.

In a few instances, grizzly bears have been observed behaving in a quite sociable way. Rob Wielgus, who undertook grizzly bear research in Alberta's Kananaskis area, told us a story about a female grizzly bear that kept in touch with her mother and daughters. "An old matriarch, whom I called 'The Grandmother,' was living in one area and had a yearling cub with her. She had two daughters, both four-and-a-half years old, who had established home ranges pretty much next to their mother's, but quite a distance away. One of the daughters had a brand-new cub. They all lived in different home ranges, but adjacent to one another's. The daughter with the new cub traveled into her sister's home range, picked up her sister, and then the three of them traveled back to 'The Grandmother' and met with her and her yearling cub. Three generations of those females were together for a period of about a week or two."

Scientists have not observed this phenomenon very often. As

Rob went on to say, one has to wonder what those bears were doing. He thinks "it was socialization, like a big female clan. You have grandmothers and mothers and aunts and nieces, so it would make sense that they would introduce themselves to each other and essentially maintain a genetic social harmony."

These characteristics of grizzly bears indicate that they are not simply slavering predators that lick their chops whenever they see a hiker. Nor are they stupid, uninteresting stomachs on four legs, eating their way through every day and staying clear of their neighbors. On the contrary, grizzlies appear to care about their children, they seem to know their relatives, and they are very intelligent — all traits we admire in humans as well.

Grizzly bears are different in one very important respect, however. They are the slowest-reproducing members of the large carnivore family in North America. Since they cannot replace their numbers quickly, they are very vulnerable to being overexploited by people. Female grizzlies do not reach sexual maturity until they are between four and eight years old, and males at five to ten. Females breed only once every two years, and at three- to five-year intervals in some areas. When they do produce litters, the average size is just two cubs, sometimes only one. Therefore, grizzlies not only reproduce slowly, they also recover very slowly from any factor that decreases their numbers. This natural fact about grizzly bears is obviously important when considering conservation measures.

Male grizzly bears do not take part in rearing the cubs. In fact, males have been frequently observed killing very young bears, which indicates that popular movies depicting old male bears "adopting" orphan cubs are highly fanciful. Since this killing of cubs by male grizzlies may be a natural population-regulation mechanism, it's not clear what can or should be done about it, if anything. For conservation purposes, it highlights again the importance of making sure female bears are not killed. This fact has led Rob Wielgus and other researchers to study sex ratios — how many males versus females there are — in certain grizzly bear populations. His concern is that, if the sex ratio significantly favors male bears, then the grizzly bear population could be limited as a result of low cub survival. Rob has looked closely at this factor in the Kananaskis area of Alberta, and speculates that the imbalance of males versus females there may have been human-caused. Some time ago, people killed large numbers of grizzly bears in this area. Since male bears travel over larger areas than females, and are dominant, they are often the first ones to recolonize an area. Such may have been the case in Kananaskis. More male bears appear to have moved there than females. Continued killing of adult male bears by humans allowed high rates of immigration by subadult male bears, which maintained the skewed sex-ratio toward male bears. That resulted in a population that is recovering very slowly, or it may fail to recover at all.

Reproductive rates for grizzlies in food-rich habitats are generally higher than rates in poor-quality areas. For example, coastal grizzlies reproduce more quickly than bears in the Arctic or in some interior regions, because the coastal bear's vegetarian diet is enhanced by salmon, an especially rich food for grizzly bears. Compared to smaller carnivores such as wolves, however, all grizzlies, even where the food supply is good, have a low reproductive capability.

Although the grizzly bear's low reproductive rate is a conservation concern in its own right, it is compounded by the fact that each grizzly needs a very large home range — 73 to 86 square kilometers (or 28 to 33 sq. mi.) in the Yukon and 179 to 1,183 square kilometers (or 70 to 457 sq. mi.) in Alberta. Fred Bunnell brings the point home. "Female grizzlies reproduce maybe every four years, so the male, if he wants to have a crack at it every year, has to have at least three female home ranges in his range. This means his range is going to be a lot bigger than the average female's. They're roaming considerably. It's darn near *space* as much as habitat that's important for grizzlies." Because these bears tend to be naturally spread out in low numbers over extensive areas, removing just one or two bears can have a widespread biological impact.

Grizzlies are omnivorous, which means they eat both plant life and meat. Their diet varies with the season, but, on the plant side, it includes berries, nuts, sedges, grasses, forbs, roots, bulbs, tubers, corms, shoots, leaves, flowers, fruits, and stems. They obtain such foods by digging underground when necessary, by grazing, and by chewing directly on fruit-bearing bushes. On the meat side, grizzly bears eat ants, bees, beetles, and even moths. Most of these they find by ripping apart rotten stumps and logs with their powerful paws, which have very long, strong claws. Other sources of meat include mice, ground squirrels, beaver, marmots, moose, deer, elk, caribou, mountain sheep, mountain goats, salmon, and carrion of all types.

Small mammals are dug out of burrows, and larger mammals are caught after short chases or quick ambushes. However, Charles Jonkel makes an interesting comment about this dreaded animal's hunting skills: "Most grizzly bears don't even know how to catch elk, deer, and such. They can become very good predators of those animals, but most of them don't have the foggiest idea how to do it and spend their time out eating flowers, nuts, and berries!"

Rosemarie and Pat Keough, wilderness photographers and authors, describe a grizzly bear that tried and missed in the Nahanni River region of the Northwest Territories. It happened in the early morning. An uninitiated film crew had emerged from the dense bush of an island in the river to see a cow moose and her calf swimming right out in front of them. Both moose were glancing back at the film crew, looking quite agitated. Then, "roaring out of the bush right beside the film crew, literally exploding out of the bush, came this grizzly bear! It burst out, charged right past them, tore across the gravel bar, and launched itself straight into the air at the moose calf, missing by inches!" The cow moose was able to nudge her calf downstream. While they swam away, the grizzly got caught in the main current and was swept downstream. Meanwhile, the panic-stricken film crew, whom the Keoughs describe as "city greenhorns," had visions of an upset, hungry grizzly bear swimming back and coming after them! "They were so scared that, within half an hour, they'd packed up their whole camp."

Grizzlies don't always miss, however. In fact, calves or fawns of various wild ungulates (hoofed mammals) can be particularly vulnerable to grizzly bear predation. In some years, and in some places, for example, the Yukon, the number of these young animals taken by grizzlies may even exceed the number taken by wolves. The result, of course, has been calls for grizzly-control programs, hand-in-hand with wolf reduction, based on the assumption that killing more grizzlies and wolves will save more elk, moose, caribou, wild sheep, and goats for humans to hunt.

Grizzlies also come into conflict with people by preying on cattle, sheep, and pigs, and by eating many crops we grow — bees and honey, carrots, corn, strawberries, raspberries, plums, and pears. These further conflicts with people lead to both legal and illegal killing of "problem" grizzly bears, creating a conservation concern. Alton Harestad, from Simon Fraser University in British Columbia, and many other carnivore biologists have noted the impact of agriculture in particular on grizzly bears: "At a broad level, if you look at a map outlining the distribution of grizzly bears in British Columbia and another map showing the distribution of cattle ranching, you might as well have cut one out of the other because there's very little overlap. It's quite impressive, actually." Obviously, people are winning when the grizzly gets in their way.

Finally, grizzly bears will feed intensively on garbage, as will black and polar bears. Stephen Herrero, in his book, *Bear Attacks: Their Causes and Avoidance*, notes that grizzlies in Banff National Park have been known to travel over 40 kilometers (25 mi.) from wilderness backcountry to feed on garbage. In Yellowstone National Park, up to seventy bears per night have been observed feeding in the dump. In the 1930s, park officials used to set up bleachers and put on "bear-feeding shows" for park visitors. Although this "habituation" to a garbage diet has made bears relatively easy to observe by people, it causes problems when people become careless around the animals. On a smaller scale, improper disposal of garbage by road crews, oil drillers, miners, construction camps, and wilderness backpackers has also caused fatal problems for both people and grizzlies.

It is not easy to make the same ecological argument for conserving grizzlies that can be made for wolves, namely, that, as predators, they play a key role in regulating prey populations. Grizzly bears eat such a vast array of foods that, with a few exceptions, they don't appear to have a major ecological impact on any particular one of them. Nevertheless, it should not be assumed that the ecological function of any animal is completely understood, let alone an animal as sophisticated as the grizzly. It has, after all, managed to survive at the top of the food chain quite nicely when not disturbed by people. Where grizzly bears have been disturbed to the point of elimination, there is no doubt that humans have drastically changed and impoverished the ecosystem as a whole. In the prairies, for example, not only was the bison eliminated, but also the bison's chief predators — the wolf and grizzly. In this respect, the real prairie has been lost forever.

However, the value of grizzlies extends beyond their ecological function. Ian McTaggart-Cowan, a respected authority on all Canadian wildlife, says that, in his view, "we want to have them because they're an intrinsic and very dramatic part of our native wildlife.

Grizzlies probably spread over here at the time that all the mammals were wandering across the Bering Strait when it was closed, and they have been part of human evolution, and part of the evolution of animals on this continent, ever since. It would be disastrous to get rid of them."

Stephen Herrero says, "One of my main interests in grizzlies is spiritual. I mean this in the broadest sense of what the grizzly bear represents regarding wilderness and the way in which people can interact with wild animals and an intact community of life in wilderness environments. I find power for myself in being part of those environments, and the grizzly bear defines that experience for me."

Perhaps this spiritual thread runs deeper through society than many of us realize. Charles Jonkel uses Missoula, Montana, as an example. Although its human population is only about 30,000, there are nearly thirty different stores, clubs, or sports teams named after grizzly bears, including "Grizzly Grocery," "Grizzly Auto," "Silvertip Lounge," and "The Montana Grizzlies." For Jonkel, this is further testimony to the enormous power of the bear. He believes bears provide a window into many broader problems we're experiencing in the world, and help us understand the environment and ourselves. North American Indians, and other cultures, used to have what they called "bear honorings." Says Charles Jonkel, "I think we need to do some similar things, only modern-day bear honorings, and understand why we're doing it."

Clearly, the grizzly has captivated the psyche of individuals and cultures. It continues to have a powerful effect on modern society, particularly in its role as a symbol of a sadly vanishing natural heritage.

WHAT HAS HAPPENED TO GRIZZLIES?

In his 1990 study, *The Status and Conservation of the Bears of the World*, Christopher Servheen reports that, worldwide, the brown bear, of which the grizzly bear is a subspecies, has lost more than 50 percent of its range and numbers since the mid-1800s "due to human intolerance." He concludes that the future of the species worldwide "can only be assured in the northeastern and northwestern Soviet Union, Alaska and Canada."

The grizzly bear has been eliminated from 99 percent of its range

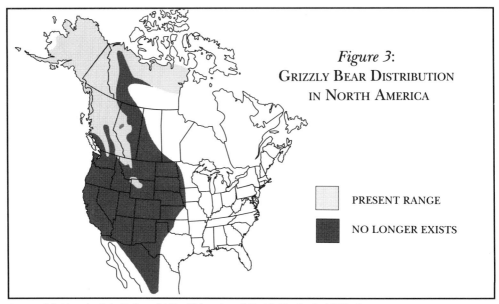

Figure 3:
GRIZZLY BEAR DISTRIBUTION
IN NORTH AMERICA

PRESENT RANGE

NO LONGER EXISTS

Source: M. Novak, J.A. Baker, M.E. Obbard and B. Malloch, © 1987, Queen's Printer for Ontario (Toronto: Ontario Ministry of Natural Resources and Ontario Trappers Association), p. 458.

in the lower forty-eight United States, reduced from an estimated population of 100,000 animals in the 1850s to only 700 to 900 animals today (see Figure 3). These last remaining bears represent just six major subpopulations, which are almost entirely confined to parks. Four of these subpopulations extend from the northern United States across the border into Canada. The Mexican grizzly bear is now regarded as highly endangered and likely extinct. Furthermore, large portions of the remaining area now occupied by grizzlies in North America are poor habitat where the bears' reproductive rates are low. Consequently, remaining populations are extremely vulnerable to overkilling by people and other disruptive factors.

Aldo Leopold witnessed the fate of the grizzly in the United States over his lifetime, and noted in *A Sand County Almanac:* "In 1909, when I first saw the West, there were grizzlies in every major mountain pass, but you could travel for months without seeing a conservation officer. Today [1949], there is some kind of conservation officer behind every bush, yet as wildlife bureaus grow, our most magnificent mammal retreats steadily toward the Canadian border. Only five States have any at all. There seems to be a tacit assumption that if grizzlies survive in Canada and Alaska, that is good enough. It is not good enough for me. Relegating grizzlies to Alaska is about like relegating happiness to heaven; one may never get there."

What has happened to grizzlies in the United States should be instructive to Canadians. John Weaver, a former bear specialist with the U.S. Forest Service, describes a 1923 map of grizzly bear distribution in the United States: "What that map showed was a classic, fragmented continental population of grizzly bears, broken into small isolated pockets in Utah and Colorado, and there were still a couple in Arizona and Oregon. If you compare that 1923 map with 1975 maps, you can see that only the largest 'islands,' such as Yellowstone, that had some refuge status, or those that were the tip of a population coming down from Canada, such as in Glacier National Park, those were the only ones that survived. The other ones that were more isolated and smaller in size have blinked out."

Is this a forerunner of what could happen in Canada? Wayne McCrory, a biologist who studies bear populations in British Columbia parks, thinks it may be: "To me, the process of extinction is long-term. It involves, in some places, gradual population elimination, which leads to localized extinctions, which leads to range fragmentation, which then leads to further problems. When we look at the time frames of extinctions, then I think the Canadian public is under a great misconception to think we're doing okay. We're just under the illusion that we're doing okay because we haven't wiped out our grizzlies the way they have in the lower forty-eight States. We don't realize that some subpopulations, in some areas, are already on their way out."

In Canada, grizzly bears now occupy just over half of their historic range (see Figure 3). They are found in parts of the mountain ranges of British Columbia from the coast to the eastern slopes of the Rocky Mountains in Alberta; in the boreal forest of the Swan Hills in Alberta; in the Yukon Territory; and in the mountains west of the Mackenzie River in the Northwest Territories. A barren-ground or tundra grizzly population appears to exist in the Arctic above the treeline in the northern Yukon and in the northern and eastern Mackenzie District and central Keewatin District of the Northwest Territories across to Hudson Bay. As mentioned before, grizzlies have been extirpated (meaning they are now extinct) from the prairies and also from a pocket in south-central British Columbia.

Table 2:
POPULATION AND HARVEST STATISTICS FOR BROWN/GRIZZLY BEARS

State/Province	Population Estimate	Reported Harvest	Status	Trend
NW Territories	4,000–5,000	60–80	Game	Stable
Yukon Territory	6,000–7,000	115	Game	Varies
Alberta	780	42*	Game	Declining
British Columbia	10,000	403	Game	Varies
Alaska	32,000–43,000	1,212	Game	Varies
Montana	500–700	14	Threatened and Game	Varies
Wyoming	200	5	Threatened	Increasing
Idaho	20–30	0	Threatened	?
Washington	10–20	0	Threatened	?

*Note, the average in 1988-90 was 10 per year.
Source: Christopher Servheen, *The Status and Conservation of the Bears of the World.*
Eighth International Conference on Bear Research and Management. Monograph
Series No. 2. (International Association for Bear Research and Management, 1990),
p. 20.

They may have existed on the tundra east of Hudson Bay, but if they
did, they're gone now.

Table 2 gives Chris Servheen's estimates for North American
grizzly bear populations, plus the number of grizzly bears killed
annually in each jurisdiction, with some refinement of the Canadian
numbers provided to WWF since his report was published. Servheen's
numbers indicate the estimated total Canadian grizzly bear population to be between 20,780 and 22,780 bears. Although he settled
for an estimate of 10,000 bears in British Columbia alone, other estimates have ranged from 6,000 to 12,500. This unacceptably wide
variance has been the topic of much recent debate, and urgently
needs to be improved. The Alberta grizzly bear population in 1990
was thought to be 789 individuals, with 574 on provincial lands and
215 in national parks. It is very difficult to provide an overall population estimate for barren-ground grizzlies, although in 1988 the
Government of the Northwest Territories produced a status report
citing densities of one bear per 200 to 262 square kilometers (78 to
102 sq. mi.). Servheen guessed the Arctic numbers to be approximately 4,000 to 5,000.

The 1991 COSEWIC status report on grizzlies divided Canada into fourteen "Grizzly Bear Zones" and estimated the total grizzly population to be 25,310 bears. Since this way of classifying the grizzly bear population is not directly comparable with Servheen's in Table 2, it is difficult to judge exactly where and why the two estimates vary, although it appears the COSEWIC report was estimating a higher number for the Yukon and British Columbia.

In round numbers, the total grizzly bear population in Canada is likely between 20,000 and 25,000 animals, of which British Columbia may have half, the Northwest Territories and the Yukon more or less equally sharing the other half, leaving Alberta with less than 5 percent. These are the jurisdictions where the grizzly bear's fate will be determined in this country.

CURRENT THREATS TO GRIZZLIES

There is intense debate about which threats to grizzly bears are the most important. One side argues that it is direct killing by humans, that grizzly bears are fairly adaptable to habitat changes, and that all the habitat preservation in the world will do grizzlies little good if people continue to overkill them. Another side says habitat preservation is the key, and that all the anti-killing sentiment in the world will do the grizzly bear little good if we don't maintain a natural system to meet its requirements. A third group is concerned with the practical problem of not knowing how much habitat disturbance grizzlies can withstand before they are displaced. All sides appear to have a common enemy, namely, roads. Roads provide access to wilderness backcountry, disturbing wildlife habitat, displacing bears, and opening areas up for human killing of grizzlies.

For the purposes of this book, we have taken the view that each of these schools of thought has obviously identified serious threats to the grizzly bear, and that overkilling, loss of habitat, and displacement must all be addressed. Arguing that one is more important than the other, when to do so could lead to neglecting a bona fide threat to grizzlies, is, at best, a waste of time and, at worst, irresponsible.

Given the slow rate at which they replace their numbers, determining how many grizzly bears can be killed by people without causing a grizzly bear population to go into decline is of great concern. It must be remembered that bears lost through human killing are additional to those lost through natural causes of death. One thing is certain: the numbers of grizzly bears killed by people must always be conservative. If nothing else, the bear's biology demands it. The 1991 COSEWIC report states that "a 4 percent total harvest rate is the maximum sustainable harvest for grizzly bear populations." Based on this assessment, on other references in the scientific literature, and on advice from bear managers, WWF is suggesting that, to be safe, no more than 3 percent of a grizzly bear population should be killed by legal hunting, and when all human causes of mortality are taken into consideration, under no circumstances should the total number of bears killed by people exceed 4 percent. We recognize that it is risky to suggest such a general guideline because some grizzly bear populations reproduce more slowly than others. These percentages, therefore, may be too high or permissive in some cases. Also, it is almost impossible to know exactly how many bears there are in a grizzly population, and how many are actually being killed by people.

Although government wildlife managers have some control over the number of bears killed legally, especially through sport hunting, they have less control over the number of animals killed as nuisance bears, and very little control over the number of bears poached or otherwise illegally killed. For example, using the numbers in Servheen's report (Table 2), it would appear that, in one year, more than 640 grizzlies were "harvested" by sport hunting out of a Canadian population of 20,780 to 22,780 bears. Therefore, the total number of bears killed this way would already be in the range of 3 percent of the estimated total population in Canada (and over 4 percent in British Columbia, according to Servheen). But how many more bears were killed as problem bears? How many more were poached? How many cubs may have been killed, or orphaned and not survived, as a result of their mother's being killed by any of the above means?

Of the twelve Grizzly Bear Zones in Canada where grizzlies still exist (they are extinct in two additional zones), the 1991 COSEWIC report states that the harvest rate exceeds 4 percent in five zones. And the report notes that this assessment is based on underestimating non-hunting kills. It concludes, "Trends that are evident at this scale suggest very strongly that local populations are being over-harvested. An examination of the mortality, including non-hunting kills, in local grizzly bear populations is imperative."

Some experts believe that the unreported, and therefore unknown, kill of grizzly bears in Canada exceeds the known kill; some estimate it at about equal; some think it is less. The COSEWIC report assumes it is 50 percent of the known or reported kill, but acknowledges that that is likely an underestimate. The truth is: we simply don't know and, because we don't know, we should err on the side of conservation. Whether we use Servheen's numbers, or those from the COSEWIC status report, it is clear that the number of grizzly bears killed each year in Canada by people through all means is already pushing, and in some cases exceeding, the 4 percent maximum. Consequently, the overall number of grizzlies taken out of the Canadian population every year is of serious conservation concern.

Rob Wielgus, a bear researcher, spends a great deal of time in the field, and has witnessed firsthand the threats bears face: "People shooting bears, either on purpose or because they are justifiably afraid, is a real problem. Especially when you have these small populations, if you shoot even one or two females, sometimes that can precipitate a decline." It is no wonder that many grizzly bear biologists agree that the three main threats to grizzly bears in Canada are, "lead, lead, and more lead." As wildlife biologist Bruce McLellan says, "They're an extremely adaptable animal, behaviorially adaptable, but they're not adaptable to being shot."

The displacement/disturbance aspect of the debate is quite serious, as well. Bears and people come "nose to nose" in competition for habitat because their needs from a natural area are very similar to ours. Areas that are good for agriculture for people also produce good foods for bears. Similarly, areas that produce the best trees for human use are often the best bear habitats. These are usually sites with productive soil, low elevation, plenty of moisture, and a warm climate. As Charles Jonkel puts it, "That's why it [a site] produces agricultural crops or trees, but that's also why it's particularly important to bears."

Jonkel elaborates on this point about competition over habitat between people and bears with an interesting analogy: "You have a saucer full of marbles and, at some point, it's full. One will roll out if you put another one in. It doesn't matter what color those marbles are. If you take bears and people as being the marbles, it doesn't matter if the marble is colored like people or colored like grizzly bears; when the saucer is full of marbles, it's full. If those marbles in the saucer are mostly people-colored marbles, then there's no room for the bears. And that's what's going on right now, at a terrible rate."

A great deal of work is currently being done on the habitat needs of grizzly bears in an effort to understand the complex range of habitats required by these animals for activities such as eating, sleeping, denning, and traveling during different seasons, day or night. Ian McTaggart-Cowan says, "Grizzlies have to have a prodigious area in order to sew together the little elements that help them make a life."

Just understanding a grizzly bear's movements for food is a difficult undertaking. Over the course of a year, it may be eating skunk cabbages and sedges along river valleys, then searching out berry crops at intermediate elevations, and rooting out ground squirrels or marmots in higher mountain country. It may be sleeping in lowland alder thickets or up on high alpine slopes during the day; denning in a mountain cave, a windfall, or a hollow tree during winter; then returning to ancient trails along river valleys or estuaries for spring and fall salmon runs. And, of course, all these special foods and places vary considerably among the coastal, mountain, boreal, and tundra regions where grizzlies are found.

In addition to food requirements, grizzly bears use special habitats for other reasons as well. In the moist valleys of British Columbia, grizzlies have ancient trails among the old-growth trees that tower above the salmon rivers. Fred Bunnell describes a trait unique to these bears called "step marking": "In some habitat types, there are little trails where they've stepped in the same spot over and over. Each footprint is sunk down into the ground, maybe several centimeters deep. It takes considerable effort, and they're obviously doing this purposefully. Everybody seems to know about bears marking trees, but it's this step marking that's quite puzzling. They're telling the other bears something, God knows what."

These diverse grizzly bear-habitat requirements have led to complicated systems of wildlife-habitat classification, where the different habitat types are evaluated for their relative importance in terms of use by the bears. The purpose of the exercise is to judge what the effects might be on bears if some of these habitats were disturbed, particularly by industrial activities such as logging, road building, mining, or flooding for hydroelectric projects, but also by recreational developments such as ski resorts, campgrounds, and backcountry trails. Salmon rivers, estuaries, and other sensitive grizzly bear feeding areas, can be destroyed or degraded by logging. However, berry crops may be enhanced by forest-industry clearcuts or recent burns. Noise from construction, mining, truck traffic, and low-flying aircraft can also alter bear behavior, resulting in disturbance of individuals or family groups. However, it is often not just one activity — one pipeline, one seismic line, one camp — that displaces grizzly bears. It's the combined effect of a number of them occurring over a number of years that results in fewer bears. Although some of these activities can be quite "non-consumptive," such as hiking trails through national parks, to avoid the stress of encountering people time after time, grizzlies will leave the area in favor of another where there is less human traffic, if such an area exists.

Transportation corridors or cutlines have been known to attract grizzly bears; they will travel along such openings and even feed on the new vegetation and berries growing there. Unfortunately, the result is often vehicle-killed bears, or changed use of formerly undisturbed natural habitats. Most important, roads provide access into once-remote backcountry, making grizzlies available for increased legal and illegal hunting. Says Rob Wielgus, "It's not roads that kill bears, it's people traveling on the roads with guns that kill bears." When asked what key steps need to be taken to conserve grizzlies, Ian McTaggart-Cowan replied, "Access is the single most important reason that bears get killed. I would like to see steps taken that make sure transportation corridors do not go through the areas of essential range in the large areas we must set aside for the preservation of large carnivores."

How much habitat disturbance grizzlies can withstand is not yet well understood. They may be more or less sensitive than we have thought. In many respects, however, we are experimenting irresponsibly by disturbing habitat before we know what the effects may be. As Charles Jonkel points out, outdoor human activities concentrate on grizzly bear habitat because we seem to want much the same things out of the land as the grizzly does. As a result, Jonkel concludes, "If you manage bear habitat well, then you're managing people habitat well. What we do with bears has a lot to do with how we treat ourselves. If you want to understand how a bear uses habitat in an area occupied by people, imagine how you would use it if you were a fugitive. Then you'll understand a lot more about bear habitat."

McTaggart-Cowan suggests that human settlement, with the access and other demands that come with it, has been disastrous for grizzlies. Fred Hovey, grizzly researcher in the Flathead Valley drainage area of British Columbia, Alberta, and Montana agrees: "Grizzlies are the first ones to go. If you look at a map of North America, you won't find much overlap between areas of human settlement and grizzly bear range. And that's especially true of agricultural development. We have large areas of ranching that don't seem to have any grizzly bears."

When people and grizzlies interact, the bears are often perceived as a "problem." They invade farms and outdoor camps, particularly where food is not properly stored, posing a threat to settlers, work crews, or hikers. However, usually the real threat results from improper human behavior, or from not understanding why a bear is "hanging around." This situation leads to a bear being labelled a "nuisance bear," and, shortly thereafter, it usually becomes a dead bear. This situation even occurs in our national and provincial parks, where grizzly bears and other wildlife should be protected.

Rob Wielgus's studies have shed some interesting light on bear behavior that humans may interpret as threatening: "One summer I got a call, and we helicoptered in and rescued this guy from what he thought was a dangerous encounter. It turned out that a female bear was hanging around the guy because she was being harassed by a male bear. She had some cubs with her, and was hanging around a pumping station because the male bear wouldn't come in there. Then the male bear did finally come in, and there was a fight between the bears. The female was injured and, eventually, at least one of her cubs was killed by the male. But what could have happened is that the female could have been shot as a nuisance bear, and there would have been no reason for it. Fortunately, no one had a gun. It's just a question of understanding the behavior of grizzly bears, in this case the female, and why she was hanging around. It's not because she wanted to eat the guy; she was using him essentially as protection for her cubs. If he'd had a gun, she would probably have been killed."

Charles Jonkel tells a poignant story about the effects on bears of displacement and disturbance, combined with human ignorance. It happened in the Flathead Valley in Montana, a very rich agricultural area, with high-density farming. There are both grizzly and black bears in the adjacent mountains, and, when they come down for food in the spring, they don't have to travel far to be in farmland. The result is an unusual mixture of bears and people at close quarters. The bears are out in the wheat fields, among the cows and sheep, and in people's gardens and cherry and apple orchards. Not surprisingly, the bears are attracted to the enormous amount of food on the valley floor.

A small, elderly woman named Millie Moran has some fish ponds on the edge of this forest/farm habitat. Millie lives alone, with a little three-foot-high white picket fence around her yard, a mowed lawn, and apple trees in her garden. There are bear trails right next to her fence, and Jonkel has seen as many as nine grizzlies and four or five black bears in Millie's "little chunk of woods." Says Jonkel, "I've asked Millie, 'How could you live right here with all these bears?' and she says, 'Well, it's easy. They stay on that side of the picket fence, and I stay on this side.'"

One day, some pheasant hunters came through, having been told there were a lot of pheasants in the woods. They hadn't been told there were bears there as well. Millie tried to warn them, but they ignored her and cut back into the woods, right where the bears were. Says Jonkel, "The bears had moved to the far edge of the woods to get away from the people. They didn't want to run across open fields in the middle of the day, so they stayed under cover. The next thing the hunters knew, they were up to their necks in grizzly bears who were trying to get back into the main part of the woods. These guys ended up killing three of the grizzlies, thinking the bears were attacking. But all the bears were trying to do was to get back into the bigger part of the woods. So help me, for five years there wasn't a black bear or a grizzly bear using those woods! This last summer, for the first time since, a single black bear was using the woods."

In addition to these problems, a new threat has surfaced for North American bears. There is sketchy but disturbing evidence that grizzly and, certainly, black bear gall bladders (for medicine) and paws (for soup) are becoming part of the international traffic in animal parts. Since black bears are likely more affected by this situation simply because they are more abundant, international trade in animal parts is discussed in more depth in the next chapter. Grizzly bear teeth and claws are also traded, both legally and illegally, for use as jewelry.

Poaching of grizzly bears and other wildlife in national and provincial parks has also become a major conservation concern. Grizzly bear heads and hides taken in this way are sold illegally as trophy mounts. As a result of recent revisions to the National Parks Act, poaching in a Canadian national park is now punishable by a maximum $250,000 fine. However, poaching is often a sophisticated business, and usually the culprits are not caught. Even if a hunter who has illegally killed a grizzly bear is arrested and successfully charged, it still means one less grizzly for the bear population.

Sometimes, a bear kill falls into a "gray area." It's not poaching, it's not elimination of a nuisance bear in a national or provincial park, and it may not even occur outside a legal hunting season. Rob Wielgus describes an incident in this "gray area" category. It is the story of an elk hunter "who killed an adult female grizzly because she claimed his elk! He'd shot an elk, took half of it out, and came back the next day to get the rest when he saw a female grizzly with two cubs feeding on the elk. Rather than saying, 'Okay, that female gets half of my elk,' he pushed the incident and started firing rounds of bullets over their heads, coming closer and closer until, finally, the female bear charged him, and he shot her. So, the mother was killed, and chances are that the cubs died that winter because she hadn't dug a den for them yet." The hunter was absolved. "It was called 'self-defense.' That happens a lot."

These examples clearly show that grizzlies have been under assault through direct killing, through habitat loss or alteration, and through displacement, especially when they are disturbed as a result of increased human access to their wilderness home. These threats continue, as do grizzly bear declines. Ian McTaggart-Cowan wrote, in *Endangered Spaces*, "There are no wildlife refuges . . . where wildlife habitats are destroyed, the individuals that occupied them are no more. There is no elsewhere! Increasing numbers of people are determined that the powerful carnivores shall survive. But these creatures will only do so if we resolve that they shall, and are prepared to alter our behaviour in ways that will leave them the habitat and resources they must have."

BLUEPRINT FOR SURVIVAL

The conservation measures outlined in Chapter 8, which cover all large carnivores in Canada, are important for the grizzly as well. In addition, however, a number of specific steps apply to grizzly bears in particular.

1. *Conservation Status in the United States and Canada*
   Grizzly bears in the United States are formally listed as a "threatened species" and are therefore subject to protection under the federal Endangered Species Act. Therefore, the killing of grizzlies is illegal, unless authorized, as in the case of dealing with nuisance bears. Illegal killing of grizzly bears is punishable by federal fines of up to $20,000 and five years in prison. Under the federal Grizzly Bear Recovery Plan, individual subpopulations can be removed from the Endangered Species Act if and when population recovery targets are achieved. Unlike the case for wolves, however, there are no plans in the United States to reintroduce grizzly bears into areas where they formerly existed.

In Canada, a new national status report on grizzly bears was prepared for the Committee on the Status of Endangered Wildlife in Canada (COSEWIC) in 1991. This report was an update of a working paper prepared in 1979. The 1991 status report came to some solemn conclusions. It identified fourteen Grizzly Bear Zones in Canada. Of the fourteen zones, the COSEWIC report concluded that grizzlies are "vulnerable" in seven zones, "threatened" in one, and "extinct" in two. It suggested that grizzlies might be considered "secure" in only four of the fourteen zones, but cautioned that "secure status does not imply that conservative management of grizzly bears and their habitats is unnecessary." After the recommendations of this status report were considered at a full meeting of COSEWIC in April 1991, the plains grizzly bear was officially designated as "extirpated," and all other grizzly bear populations in Canada were designated as "vulnerable." (See Appendix B for COSEWIC definitions.)

As WWF participated in and voted at the COSEWIC meeting where these decisions were made, we can report that, if anything, these designations may still understate the status of grizzly bears in Canada. The population figures upon which these decisions were made are uncertain, and the future "upgrading" of some subpopulations to the "threatened" category is likely, especially given the fact that grizzlies are being overharvested in five zones, primarily through the Rocky Mountain range. Therefore, WWF is recommending that the existing status decisions regarding grizzly bears by COSEWIC be carefully monitored every year, so that they can be updated as necessary for the conservation of grizzly bear subpopulations at risk.

2. *Direct Killing by Humans*
So far, no subpopulation of grizzly bears has been formally classified as "threatened" or "endangered" in Canada. Should this occur, however, the hunting of such a population should not be permitted by provincial or territorial governments. In addition, if the hunting kill on any subpopulation exceeds 3 percent of the estimated grizzly bear population, an immediate moratorium should be placed on sport hunting until or unless reduced levels of killing can be determined, and regulated to within safe levels.

Quotas and regulations regarding the legal killing of grizzlies must always be extremely conservative, especially when there are other people-caused sources of killing, such as poaching and nuisance-bear removals. In the past, very few grizzly bear populations have been hunted without population declines resulting. However, various management techniques have been tried, some successfully. One allows only a late fall hunting season, which helps select for the killing of male bears rather than females. A point system in the Yukon also shows promise because it assigns different values to bears according to age and sex. The idea is that once all the points have been used up, hunting stops.

Although the hides and meat of some large carnivores are important to indigenous peoples for subsistence purposes, the ethical and biological case for killing grizzly bears for sport or trophy hunting is becoming increasingly difficult to make. The removal of even a single grizzly bear can have a serious biological impact on a particular bear population. Grizzly bears, after all, do not function as prey species under natural conditions, and, as pointed out earlier, their reproduction reflects that fact. Furthermore, if cubs aren't present or seen, it is very difficult for most hunters to distinguish between male and female bears in the field. If the grizzly they shoot is a female, that one squeeze of the trigger eliminates a crucial breeding component of a wildlife population that is not easily, if ever, replaced. The elimination of spring bear hunts would further protect females with very young cubs.

WWF is not an anti-hunting organization, but it anticipates the day when this particular "superspecies" is respected and left alone in the wilderness setting it has come to represent. *Outdoor Canada* magazine, speaking, as it does, on behalf of sportsmen and -women across the country, is certainly not an anti-hunting publication. Yet, in the October 1990 edition, its editorial posed this question: "With all the press that the grizzly has had lately — losing its home territory and dropping numbers — why don't hunters step right up to the podium and support a ban on grizzly hunting?"

The reporting of all "defense" kills, whether for human safety or to protect property, and the reporting of all problem or nuisance grizzly bears killed must be made mandatory in all jurisdictions of Canada. The day has passed when private individuals can quietly take the law into their own hands to deal with these animals. If there is a good reason for such killing, there should be nothing to hide from wildlife officials.

3. *Habitat Loss*

Management plans to minimize human impacts on grizzlies must be instituted when considering habitat-altering activities such as road building, locating hiking trails, authorizing low-level over-flights, planning recreational-facility developments, engaging in fire suppression and forest spraying, permitting livestock to graze in alpine uplands or river valleys, constructing hydroelectric dams, logging, building housing, and licensing oil and gas development and hardrock mining.

These management plans must not simply serve as studies that allow such developments to go ahead; they must serve as assessments as to whether or not such activities should be permitted at all in grizzly bear habitat. Critical seasonal feeding areas, denning sites, and traditionally-used travel routes for grizzlies should be protected by law wherever they occur.

4. *Access*

One common factor in almost all human killing, habitat disturbance, and displacement of grizzly bears is access. If this slow-reproducing species is to be successfully conserved in wilderness habitats, first and foremost, serious thought must be given before opening up grizzly bear backcountry, particularly using roads and rough trails, which can be traveled these days by virtually any kind of vehicle. In addition, existing access roads may have to be closed or withdrawn, for example, by replanting natural vegetation along the former right of way. Unfortunately, once the crucial step of providing access has been taken, all other steps to conserve grizzlies may simply amount to "closing the barn door after the horse is gone."

Alton Harestad suggests that access is one conservation step we can do something about more easily than others: "You're not putting hundreds of people out of jobs. And it's something you can vote for today and have accomplished by tomorrow. There are habitat manipulations, which try to enhance habitat, that will probably take years, and they cost money. I just don't understand why we can't seem to do the easier kinds of things. Close access roads!"

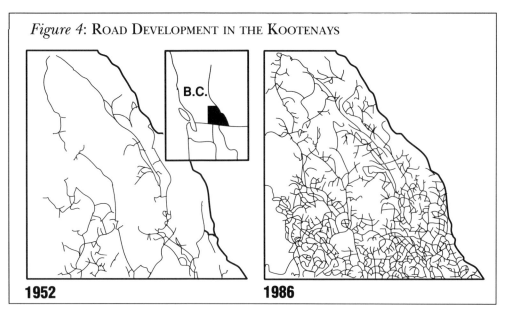

Figure 4: ROAD DEVELOPMENT IN THE KOOTENAYS

B.C.

1952

1986

Source: B.C. Wildlife Branch, Cranbrook Regional Office

Figure 4 provides a graphic example of what access can do to a formerly remote piece of Canada. As an interesting postscript to this map of the Kootenay region of British Columbia, Bruce McLellan, who worked on it, noted that many new roads have been opened up since the map was prepared.

5. *Sanctuaries*

Aldo Leopold wrote, "Saving the grizzly requires a series of large areas from which roads and livestock are excluded." No sanctuaries exist in Canada expressly to protect grizzlies, though candidate areas are known. Such reserves must be established soon before the option to do so is lost, and they must protect distinct local subpopulations in all regions. National parks and game reserves now account for only 6 to 7 percent of known grizzly bear range. Because virtually none of these were originally established for this purpose, current boundaries and management activities may not be appropriate for conserving grizzlies and other large carnivores in today's world. Even in 1949, Leopold was urging that national forests surrounding U.S. national parks be used to enlarge the parks to better accommodate grizzlies. It's time to consider similar proposals for buffer zones around our parks in Canada, while we still have the opportunity to do so.

6. *Bear Attacks and Human Safety*

With respect to preventing encounters that can result in the death of people or grizzlies, Stephen Herrero has identified the following issues: the location of trails and campgrounds and deciding what levels of use for both should be permitted; keeping human food and garbage away from bears; managing problem bears, for example, by trapping, drugging, and relocating; and communicating better information about bears.

Time and again, people who have spent their working lives in grizzly bear country emphasize that, although grizzlies can be dangerous and must be respected, the notion that they are ferocious killers of humans is simply not factual. It bothers Stephen Herrero to see this image of ferocity exaggerated: "It certainly is true that a grizzly bear female with young will occasionally charge and injure a person who is at close range. But studies have shown that, even when grizzly bears have been shot and wounded, about three-quarters of the time they just try to get out of the way or go into cover. They don't automatically charge, even when you meet them at close range on a trail. The normal response, even under these circumstances, is for the bear to avoid contact and go the other way." Wayne McCrory adds, "People's fears are constantly reinforced because every mauling or fatality in Canada gets nationwide media attention. But of the thousands of times grizzlies move out of people's way to avoid a problem — that's never recorded." This excessive fear of grizzlies can unnecessarily ruin our enjoyment of wilderness. "I've talked to people who are so afraid of grizzly bears that they don't want to go hiking," says Rob Wielgus. "Or when they *do* go hiking, they have this constant fear that they might bump into a bear. And their hiking experience is not nearly as enjoyable as it could be because they have this unreasonable fear. So, it doesn't help human beings and it doesn't help the bears."

7. *Government Cooperation*

Representatives from the four jurisdictions in Canada with grizzly bears (the Northwest Territories, Yukon Territory, British

Columbia, and Alberta) plus Canadian Parks Service representatives, should confer annually to establish population conservation goals, and to assess whether their respective conservation efforts are assuring the long-term survival of the species and local subpopulations on a national basis. British Columbia is preparing a species-management plan, Alberta has just released its management plan for grizzlies, and the Northwest Territories has prepared a status report and management plan for the barren-ground or tundra grizzly. These plans, plus those for the Yukon, must all be completed, made public, implemented, and compared, to see whether they add up to a sufficient *national* effort to conserve grizzlies in Canada.

International cooperation with the United States through Canadian participation on the Interagency Grizzly Bear Committee is also important and should be continued. This helps coordinate efforts to conserve grizzly bear populations shared by both countries.

8. *Research*
Rob Wielgus jokes that a grizzly bear is always "smarter than your average grizzly bear researcher"; nonetheless, research needs on grizzlies are particularly important for conservation purposes. They include:
• developing better census techniques, so wildlife managers can assess whether or not some grizzly bear subpopulations are being overexploited;
• monitoring human impacts on grizzlies and their habitats, to better understand their extent and cumulative effects;
• establishing nutrition levels to better understand key bear foods for purposes of habitat protection and improvement;
• studying ways to frighten off grizzly bears, so that people can deal with problem bears without having to kill them; and
• studying bear behavior and learning in relation to humans and livestock, so that landowners can take steps to minimize problems in this area.

All of these research efforts require financial support at government, non-government, and academic levels. However, everyone can play a role in this by financially supporting such work directly through non-government organizations, and by insisting that a portion of our tax dollars to government be used for such work. There is also a need to insure that such research is carried out with independence and professionalism. Charles Jonkel lays it on the line when he says, "Bears are forever getting kicked around by bureaucrats, by politicians who make biologists into bureaucrats, and by local economic interests, whether it's loggers, farmers, ranchers, hunters, or whatever. Research and management have to be done by professionals, and have to be protected by government agencies. This is not happening. Professionals are being corrupted. When it comes to bears, there has been a lot of political influence."

The grizzly bear has come to represent a vital dimension of Canada's wildlands. Its presence is a kind of assurance that we still have large wilderness areas here. Conversely, its absence clearly indicates that we are failing to maintain parts of this planet in a wild state. As Stephen Herrero says, "If we learn to save grizzly bears, then we can also save a whole lot of other things and maintain a quality of life for human beings. That's something I'd like to be part of. So, I see carnivores as being surrogates for whole ecosystems that are functioning and working well on their own. I think if we can continue to maintain those, then we will probably survive as a society."

4.

The BLACK BEAR

When it comes to black bears, I think the most common misconception is that

they are basically pests, and that people can kill them at high rates.

FRED HOVEY, BEAR RESEARCHER, BRITISH COLUMBIA

SINCE BLACK BEARS ARE RELATIVELY NUMEROUS, THEY ARE THE TOP PREDA-
tor most of us are likely to see. In fact, the major problem with
respect to conserving these animals is the common belief that they
are so abundant that there is no need to worry about them.
However, both government and non-government experts *are* concerned about the future of black bears, even though there may be
ten times as many of them in North America as there are wolves.

BLACK BEAR FACTS AND FIGURES

The American black bear (*Ursus americanus*) is indigenous only to
North America. A creature of forested land, it is the most widely distributed of our bear species. It is also the smallest in terms of average
size, although certain large individuals may attain the size of a grizzly
bear. Adult males range in weight from 120 to 280 kilograms (265 to
620 lbs.), while adult females weigh in at 45 to 180 kilograms (100 to
400 lbs.). Black bears have a longer facial profile than grizzlies, and
their body forms a straighter line along the back, lacking the notice-
able hump on the grizzly bear's shoulders.

Not all black bears are black. Their coat color ranges from blond
to cinnamon, to light brown, to dark brown and jet black. There are
bluish-tinged black bear populations, also called "glacier bears," in
parts of coastal Alaska and northern coastal British Columbia, and
even white black bears, known as the "Spirit" or Kermode bears,
found only on certain coastal islands off British Columbia. In North
America, black-colored bears seem to be most plentiful in moist
areas, whereas brown-colored bears tend to be found in the drier
interior regions. Unlike grizzly bears, black bears often have a white
"V" on the chest, and a noticeably lighter tan area along the snout,
which contrasts with the darker nose. To add to the color confusion,
individual black bears may vary in color over the year because they
molt in the summer. Their new darker hair becomes gradually
lighter up to the next molt, especially in brown-colored bears.

In his basic text, *Mammals of Canada*, A.W.F. Banfield suggests
there are ten subspecies of black bears in Canada. One subspecies is
found only in Newfoundland, one subspecies is distributed through
virtually all of central Canada, and there are no less than eight subspecies found in British Columbia and the Yukon. Many of the
western subspecies reflect the color phases already mentioned. In

addition to the white bears (*Ursus americanus kermode*) and bluish-tinged populations (*Ursus americanus emmonsii*) in British Columbia, there are brown-colored bears (*Ursus americanus cinnamonum*) in the interior region of the province, and other important local sub-species on Vancouver Island (*Ursus americanus vancouveri*) and the Queen Charlotte Islands (*Ursus americanus carlottae*). Elsewhere in Canada, the black bears of Riding Mountain National Park in Man-itoba have been identified as exceptionally large, and Ian Stirling, Canada's polar bear authority, has reported a unique black bear population north of Saglek Fiord, Labrador, living on the tundra above the treeline — country we would normally think of as grizzly, not black bear habitat. Perhaps the grizzly bear was found here once, but now black bears are moving into their former range.

As for all other top predators, we must design conservation strate-gies to maintain these different subspecies or subpopulations. We cannot expect to conserve "the black bear" by focusing on only one or two areas in Canada. Rather, we must protect the diverse habitats that have produced diverse black bear subpopulations.

Black bears generally reproduce more quickly than polar bears or grizzly bears, but they are still slow reproducers relative to other large wildlife species. The age at which female black bears become sexually mature depends upon the quality of their environment, especially the food supply. In east-central Ontario, where food is not consistently in good supply, female black bears do not have their first litter until ages five to seven. However, on average in North America, sexual maturity occurs at three or four years of age. Litter sizes range from one to four cubs, even five, though two cubs are most common. Adult females usually reproduce every two years, but, again, depending upon food supply. If food supply or habitat quality is low, female black bears may reproduce only every three or four years, and there are examples of Canadian black bears reproducing at intervals as long as nine years!

Cub survival in the first year of life is usually quite high, averag-ing 80 percent. However, cubs of poorly fed females are more likely not to make it. As many as 50 percent may not survive in food-scarce years, and only 30 percent of cubs orphaned in the spring survive. These figures are worth keeping in mind in view of the fact that most jurisdictions in Canada still permit a spring black bear-hunting season.

Although they may be abundant, black bears are sensitive to excessive killing by humans because of their relatively low reproductive rate, especially in areas of poor-quality habitat. A typical female black bear, sexually maturing at four to five years of age and producing two cubs every two years, would produce a maximum of five or six litters, or ten to twelve cubs, over her fifteen-year life span. If anything, she might give birth to fewer cubs. These figures refer only to cubs produced; the number that actually survive to become adults would be much smaller, and, on average, only half of those would be females. Therefore, a mother black bear may be responsible for adding only a small number of new reproductive females to the bear population during her entire lifetime.

The conservation lesson here is that we must make sure that areas of prime habitat where black bears must get their food are protected to conserve healthy bear populations. As well, in poorer habitats, we must insure that pressures such as sport hunting are carefully adjusted to allow the survival of subpopulations that may not reproduce as quickly.

Although they have special habitat requirements for bedding, resting, hibernating, and even mating, the main activity that governs the location and behavior of black bears is eating. Charles Jonkel, who has studied bears extensively in Canada and the United States, explains: "The major difference between bears and most of the other species is that they den for a good part of the year, anywhere from five to seven months. This means they have only five to seven months of the year to eat. Eating is extremely important to them! They're very large animals, so they need a lot of food, and their food comes from their habitat. For a couple of the months that they're eating, the food isn't all that great, and they're just barely making it. Therefore, they generally have only two or three months out of the year to make their living, not just for those two or three months, but for the other nine or ten months as well."

About 75 percent of most bears' diet is vegetarian — twigs, leaves, catkins, shoots, buds, roots, grasses, sedges, berries, and nuts of various kinds. Usually, small animals are too fast for black bears to catch, but they will eat fish and carrion of all kinds, and calves of moose and deer in the spring when they are newly born and particularly vulnerable. Black bears also frequently feed on ants, grubs, and beetles ripped out of rotting logs, as well as crickets and grasshoppers. Crops such as strawberries, cherries, apples, oats, corn, and honey are favorite foods, too. As in the case of grizzly bears, all of these food sources may vary according to the season and region of the country where black bears are found.

Black bears are well known for feeding on human garbage. As Stephen Herrero explains in his book *Bear Attacks: Their Causes and Avoidance,* "People's food and garbage are so attractive to bears not because bears 'eat anything,' but rather, because people's food and garbage are so easily converted into calories by bears." In years when supplies of their natural foods, such as berry crops, are low, bears may seek human food and garbage more aggressively. People who live in black bear country have noticed that more bears show up in camp or in town when their natural food is in short supply. Because reproductive success for female bears is closely tied to nutrition, Herrero concludes, "We can now clearly see the basis for the attraction of female bears to garbage-food sources . . . they are able to increase their body weight and hence produce more offspring."

In planning for the conservation of black bears, we would do well to keep in mind that they don't become habituated to garbage out of some whimsical or perverse desire to become a nuisance, but out of a fundamental drive for survival and to continue their species. That helps to explain why packs, coolers, tents, boats, cars, and trucks have been literally ripped open by black bears trying to access human food.

Obviously, if these animals are destroying human property, they will come into conflict with people. But most direct black bear attacks on people are a result of human carelessness or foolishness. Such is especially true of bears in parks. Stories abound of people putting children on a bear's back for the camera, holding food up so the bear will "dance" on its hind legs, luring a bear inside a car so it sits in the driver's seat, or even smearing people's faces with honey or peanut butter to get a "bear kiss." Black bears are also harassed by people throwing rocks while the bears feed in garbage dumps, or they are approached too closely by photographers. It is difficult to blame the bears for problems that occur under such ridiculous circumstances.

In researching this book, we heard many stories not only of people watching bears, but of bears watching people. Malcolm Ramsay, of the University of Saskatchewan, told us about watching some bighorn sheep, when he noticed a black bear foraging in a distant valley. At the time, Malcolm was interested in finding out which predators were preying on the sheep. One way to do that is to look through scats (droppings) from the predator for bone or hair that would indicate what it had been eating. Malcolm made a note in his field book of the bear's location so he could go down to the site later on, to look for bear scats. "The next day I was down in the valley, on my hands and knees, going through tall grass, looking all over for bear scats. I caught a movement out of the corner of my eye, and looked up. Here was, I'm sure, the same bear about three meters away from me, standing up on his hind legs, looking down, trying to figure out what I was going to do. I suspect we were both fairly surprised! In retrospect, I found it quite humorous, though I'd done what was actually a fairly foolish thing, putting myself in a place where I knew there was a bear foraging. In fact, nothing happened. The bear didn't make any effort to act aggressively toward me."

Fred Hovey describes watching a colleague shinny up a hollowed-out tree, looking for a female black bear's den inside: "He was going up the tree on the outside, and she was going up the tree on the inside and they met at the top! He peered over just as she peered out, and they almost touched noses. He went sailing off the tree as she went sailing back down!"

As with grizzly bears, there is always the problem of stumbling into a female black bear with cubs, a situation that should usually be met by the person backing off slowly and quietly, to become as unthreatening as possible. However, researchers have found that black bear females are actually seldom very aggressive. Herrero's research shows that 90 percent of the recorded black bear-inflicted injuries are minor, whereas over half the injuries from grizzlies are more serious. "When a wild black bear suddenly encounters a person," he says, "it frequently will charge toward the person, swatting the ground with a front paw or making loud blowing noises. Although such actions may make your palms sweat and your legs shake, they are rarely followed by attack."

Bruce McLellan, a wildlife biologist with the Government of British Columbia, speculates about an interesting personal encounter he had with a black bear: "I was going through a berry field when, all of a sudden, there he was. Quite a few times he did that bluff charging and straight-legged-walk stuff. Then he finally worked himself around to where he had a tree between me and him. As soon as he hid behind that tree where I couldn't see him, he just took off running the other way as fast as he could. I almost got this feeling that it was the 'macho mood' he was in, that he didn't want to show that he was really frightened, but as soon as he broke visual contact, it gave him a chance to leave in a hurry."

None of this should be read as suggesting that black bears are harmless. In fact, Dennis Voigt, a research scientist with the Ontario Ministry of Natural Resources, and Stephen Herrero, have both speculated that, in a few isolated situations, black bears may have actually considered people as prey. This appears to have happened in rural or remote areas where black bears have relatively little exposure to people. This type of behavior is judged to be exceedingly rare, and Herrero has described steps that may be taken to avoid or even to deal with such encounters in *Bear Attacks*. Since there are steps we can take to avoid bear attacks, an essential part of conserving the species is for us all to heed information regarding common-sense precautions that can be taken to insure safe recreation and work in black bear country. Following such advice helps protect both humans and bears.

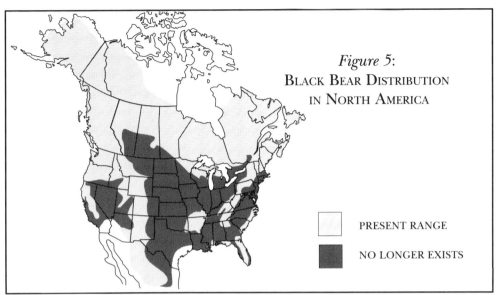

Figure 5:
BLACK BEAR DISTRIBUTION
IN NORTH AMERICA

PRESENT RANGE

NO LONGER EXISTS

Source: M. Novak, J.A. Baker, M.E. Obbard and B. Malloch, © 1987, Queen's Printer for
Ontario (Toronto: Ontario Ministry of Natural Resources and Ontario Trappers
Association), p. 444.

WHAT HAS HAPPENED TO BLACK BEARS?

Black bears generally don't like to be too far from forested land of
some kind, which they use for both food and cover. They also climb
trees for safety, to escape other predators or harassment. Their
range in North America, therefore, once included virtually all the
wooded area of the continent. Today, the black bear has disappeared from nearly all those areas where forests have been perma-
nently cleared.

Figure 5 shows the original and the current range of black bears
in North America. Their current range is about 9.8 million square
kilometers (3.8 million sq. mi.), or about 60 percent of its range in
1800. The black bear has disappeared from deforested areas in the
eastern and central United States, and from most developed southern regions of Canada. Sadly, that means that even our most abundant top predator is gone from more than half of its former range
in North America.

However, black bears in a few areas are benefiting from
changes to their habitats. Some aspects of human settlement, for
example, orchards and grain crops, may have led to increased

numbers of black bears, if sufficient tree cover also remains. And black bears have expanded their range into areas where grizzlies no longer exist, such as the Ungava Peninsula in northern Quebec. Unlike other large carnivores, black bears have been successfully transplanted to re-establish healthy populations in Arkansas, Louisiana, and other states. They now live in forty-one American states, and in a few mountainous places of northern Mexico. In Canada, they occupy about 85 percent of their former range, having been eliminated from the southern parts of virtually all the mainland provinces.

The trends regarding black bear populations in Canada are similar to those for wolves. Like the wolf, the black bear is still a relatively abundant predator, but it has already disappeared from the southern portions of its range, making way for people's intrusions into its habitat. As has been the case with wolves, bounties have been offered for killing these bears — as recently as 1961 in Ontario. Throughout Canada, trapping and sport hunting of both animals are permitted, usually with spring and fall hunting seasons for black bears and no closed seasons at all for wolves. Both predators have been persecuted as undesirable pests by livestock owners and other farmers, who have sometimes been paid compensation for livestock or crop losses proved to have been caused by bears or wolves. In British Columbia, an average of 400 black bears per year are killed by the government Wildlife Branch for damaging human property or jeopardizing human safety. Across Canada, there has also been extensive illegal killing, by shooting, poisoning, and trapping of both black bears and wolves, as some landowners and other private citizens take matters into their own hands. All of these activities and control measures have removed large numbers of both species from their Canadian populations.

Now attitudes toward the black bear and the wolf seem to be changing. Black bears have been "upgraded" to the status of "game" animals, which means they are subject to hunting regulations. Of some concern, however, is the recent dramatic increase in the hunting of these bears by non-residents of Canada through commercial guide-outfitting businesses, especially in New Brunswick, Ontario, Quebec, Saskatchewan, and Manitoba. Most of this hunting is done by American and European hunters who come to Canada solely to hunt black bears. It was considered to be

Table 3: POPULATION AND HARVEST STATISTICS FOR BLACK BEARS

Province/Territory	Population Estimate	Annual Kill
Alberta	48,700	2,800
British Columbia	63,000	3,500–4,000
Manitoba	?	1,500–1,750
New Brunswick	?	973
Newfoundland	6,000	103
Northwest Territories	?	?
Nova Scotia	3,000	500
Ontario	65,000–75,000	8,701*
Prince Edward Island	Extinct	0
Quebec	60,000	2,000
Saskatchewan	30,000	1,674
Yukon Territory	10,000	103
Totals	**285,700–295,700**	**21,854–22,604**

*1986 Statistic
Source: Christopher Servheen, *The Status and Conservation of the Bears of the World.*
Eighth International Conference on Bear Research and Management. Monograph
Series No. 2. (International Association for Bear Research and Management, 1990), p. 22.

such a problem in Saskatchewan in the 1980s that the provincial government limited the number of hunting licenses allocated to non-resident sport hunters — quite a contrast to the days when the attitude was: "Take as many bears as you want; we want to get rid of them."

With these new pressures, wildlife managers are now coming to appreciate this species's slow reproductive capability. Black bears have become scarcer in many areas, even in our national parks, as a result of extensive killing of nuisance or problem animals. Concern is also growing for black bears in poorer-quality habitats such as the Canadian Shield in eastern Canada, where the population is less able to withstand the pressure of heavy killing by people.

Despite all this, black bears remain abundant in North America, possibly numbering 500,000 animals, although this estimate could be larger or smaller by as much as 200,000. Table 3 gives Chris Servheen's population and harvest statistics for black bears in Canada, which is drawn from a larger table presenting such information for all of North America. These numbers are imprecise, because no population estimates were available for Manitoba, New Brunswick, and the Northwest Territories. Nevertheless, for general guidance, Servheen's report is useful. It indicates, for example, that Canada likely harbors over half of the North American black bear population — approximately 285,700 to 295,700 bears out of an estimated North American population of 458,410.

Table 3 further indicates that 21,854 to 22,604 black bears are killed through hunting each year, representing as much as 7 to 8 percent of the total Canadian population. In some provinces, however, the percentage of bears killed by hunting is significantly larger; in 1986, in Ontario, for example, it was 11 to 13 percent of the estimated black bear population. If Servheen's numbers are correct, this level of hunting kill appears to be nearing what is considered to be safe conservation limits, particularly given the fact that the data do not include estimates for the number of bears killed illegally. Perhaps this finding reflects the still prevalent view that these predators are vermin, akin to big rats. It is also interesting to note that about the same number of black bears are killed every year in Canada as what is estimated to be the entire grizzly bear population in this country.

CURRENT THREATS TO BLACK BEARS

A key threat to black bears is loss or degradation of their habitat. Like so many other top predators, black bears are being gradually beaten back from their original range in the south, as forested land is converted to agricultural and urban use. The result is usually permanent loss of their habitat. Industrial operations can have the same effect. Logging, in particular, can be damaging to black bears in the short term if clearcuts are too large and if access roads facilitate overkilling through increased hunting pressure. However, cutover lands that produce good berry crops can increase food supply, thereby benefiting bears, provided such areas aren't sprayed, and provided sufficient tree cover is left to accommodate other aspects of the bear's lifestyle.

Protected areas such as provincial and national parks do preserve habitat for some black bear subpopulations, although this depends on what the particular park policy is regarding hunting, and killing versus relocating nuisance bears. In general, protected areas in Canada were not established to conserve wildlife habitats for large carnivores, with the result that many important subpopulations of black bears are not protected in this way.

Sport hunting as a threat to black bears is a sensitive topic because the hunting of this animal is an important source of income for those who operate guide-outfitting businesses in the North, and because black bears are generally regarded as abundant. Nevertheless, sport hunting can definitely affect the composition of a black bear population in terms of sex and age structure, and, in areas where hunting pressure is heavy, it can seriously limit the overall bear population.

Joe Robertson, a hunter and trapper from Manitoba, has concerns about the increasing commercialization of black bear hunting there: "Although we like to hunt, we feel that the animals need more protection. We're fighting tooth and nail against the decimation of the bear population because they've commercialized it and allowed the outfitters to have so many licenses to bring in non-residents. One guide-outfitter alone has fifty bait stations that he puts out in the spring! Well, you know that's just a deadly thing! We must be alert to commercialization of wildlife. Once it's commercialized, then we're in a different ballgame."

In remote, lightly hunted areas, it is usually the male bears that are killed because they roam farther and are less cautious than females. But, as hunting pressure increases, more female bears are killed. Once adult females begin to be killed, their replacement can be slow, and it may take ten years or more for a black bear population to recover.

The more than 20,000 black bears per year that are legally killed through trapping and sport hunting in Canada is thirty-five times the number of polar bears or grizzly bears that are killed in this way. In Ontario, Quebec, Saskatchewan, and Manitoba, there is no limit on the number of black bears that may be killed through trapping. Legal spring black bear hunts, and baiting, are common throughout the country. These practices reflect the view that, in Canada, black bears are considered abundant at near "pest" levels. In contrast, in the United States, where black bear numbers have decreased, few states now permit trapping or spring hunts. The fact that so many hunters come to Canada from other countries to shoot black bears indicates what has happened to these bears elsewhere. It also indicates that Canada, with its lax hunting regulations, is regarded as one of the last remaining "motherlodes."

International trade in bear parts is a growing threat to all bears, but to black bears in particular. The trade stems primarily from the use of bear parts in traditional East Asian medicines, for the restaurant trade, and for jewelry. Bears' gall bladders, claws, feet, and teeth are in great demand.

In traditional Chinese medicine, bear gall bladder — the pharmaceutical name is "Fel Ursi" — is believed to help treat high fevers; convulsions; delirium from extensive burns; skin lesions; visual obstruction; red, painful, and swollen eyes; and swelling and pain caused by trauma, sprains, fractures, or hemorrhoids. Often, the gall bladders are dried and used as crushed powder in combination with other animal parts or herbs. China's best-known herbal handbook lists 444 animal-part substances. These medicines were brought to Japan and Korea in the fifth century. In this century, the growing popularity of traditional medicines has resulted in a significant increase in the trade of animal parts to China, Japan, and Korea from all over the world.

There is still a great deal of scientific research to be done on the efficacy of traditional East Asian medicines. Many of the substances have no known active ingredient, or, where there is an active ingredient, the medicinal effects are not well understood. Nevertheless, advertising for bear parts is now common in Canadian outdoor magazines and newspapers. One advertisement originating from Hong Kong and addressed to "Meat Inspectors, Ginseng Diggers, Hunting Guides, Hunters, Slaughtermen and Veterinary Surgeons," reads as follows:

> Here's some extra big money while you work or play (that's hunting). I buy animal by-products which you always come across but you might not know that they can bring good money.
> Bear Gall Bladders
> Price: $100 per 100 gm. (dried)
> If you hunt bears, remove the gall bladder and sun dry them until the end of the hunting season when they are 100% dried. But send in your first gall bladder as soon as it is dried to ascertain you have the right thing.

Complete instructions regarding packing, mailing, and payment were provided in this advertisement.

Andy Ackerman, a senior Wildlife Conservation Officer for the

Government of British Columbia, has reported on this trade in Canada and the United States. According to him, a large bear gall bladder, which sold to a local buyer in British Columbia for $5/gram ($2,250/lb.), could fetch up to $30,000 in Korea. Undercover agents in California have discovered bear gall bladders selling for $20/gram. That would make black bear gall bladders worth more than drugs or gold on the international market. Bear feet are sold for $2.50 to $10.00 each without claws, and $60 per foot for use in soups in restaurants. Claws are sold for $2.50 to $7.50 each. Ackerman points out that, in British Columbia, one buyer alone purchased 1,125 bear gall bladders in less than one year.

A 1990 report from Manitoba indicated that a dealer in Ontario placed an order for 5,000 gall bladders. Bear teeth in Manitoba sell for $60 each for jewelry, and skulls for $50. Some American states have found dealers shipping fresh whole dead bears, packed in ice, to Korea.

Japan has become a major importer and processor of bear parts. From 1979 to 1988, between 11,000 and 59,000 bear gall bladders were exported from China to Japan. The IUCN Bear Specialist Group indicates that, between 1978 and 1988, 681 kilograms (1,513 lbs.) of bear gall bladders were exported from India to Japan, representing as many as 8,000 dead bears. At 1990 prices, a recent shipment of 61 kilograms (134 lbs.) of gall bladders from China to Japan would have been worth almost $4 million. The annual export of bear paws from China to Japan is estimated to be about 600 kilograms (1,330 lbs.).

These are only examples of the known trade in bear parts. The unknown trade is likely even larger. This trade has already seriously endangered several Asian bear species, leading to concern about whether North American bears aren't far behind. Furthermore, because it is very difficult to distinguish a North American black bear's gall bladder from that of an endangered Asian bear, traders simply claim they are dealing in the more abundant, legally traded species in order to avoid prosecution. The high profits from "laundering" — deliberately using "look-alike" parts — have spawned a massive worldwide traffic in wildlife. The future impact of this trade will no doubt be felt most heavily by Canada's most abundant bear — the black bear — in the form of illegal killing for the Asian medicine and restaurant trade, both inside and outside the country.

This is, indeed, a very disturbing "storm cloud on the horizon" for this creature of Canada's forested wilderness.

BLUEPRINT FOR SURVIVAL

The conservation measures outlined in Chapter 8, which cover all large carnivores in Canada, are important to the future of the black bear as well. In addition, however, a number of specific steps apply to the black bear in particular.

1. *Conserving Black Bear Subpopulations*

   Because the black bear is so abundant and widespread in Canada, its range covers many different natural regions, habitat types, and ecoregions. Therefore, it is particularly important to conserve this species at the level of the unique local bear populations that have evolved in these diverse habitats. Protected areas such as parks and wildlife reserves should be located with this in mind. Conserving these local gene pools, as part of the goal of conserving overall biological diversity, should also be a guiding principle when adjusting hunting regulations or modifying logging, mining, or other industrial land uses on that part of the landscape where such activities are permitted.

2. *Hunting Regulations*

   Depending on food availability and quality, it is believed by wildlife managers that between 3 and 8 percent of a black bear population can be killed by people without causing it to go into long-term decline. Current levels of killing in Canada are certainly at the upper limit, if not exceeding it, a situation that cannot and should not continue.

   Black bears are abundant enough to withstand some hunting, but it is time to abolish or revise outdated practices that lead to an excessively large number of black bears being killed in Canada. Even though such a move will be controversial and vigorously opposed, spring black bear-hunting seasons should not be continued. Fall hunting-season dates should be adjusted to protect female bears. Because female bears usually enter their hibernation dens before males, a later fall season should result in fewer females being killed. In general, males move

Table 4: Bear Parts Trade Activity and Trends in Canada

Province	Trade activity in province (legal or not)	Legal to sell parts?	Trends in trade within the province	Parts going to out-of-province destinations?	Local use of parts within province?
Alberta	infrequent	yes	stable–growing	yes	yes
British Columbia	active	yes	growing	yes	yes
Manitoba	active	yes	growing	yes	yes
New Brunswick	active	no	unknown	yes	no
Northwest Territories	active	yes	none	yes	yes
Nova Scotia	infrequent	yes	stable	yes	yes
Ontario	active	no	growing	yes	yes
Quebec	infrequent	yes	growing	yes	yes
Yukon	active	no	growing	yes	yes

Source: Lili Sheeline, *The North American Black Bear (Ursus americanus): A Survey of Management Policies and Population Status in the U.S. and Canada,* a report submitted to the World Wildlife Fund U.S., June 1990.

around more than females. Therefore, a hunter's chances of encountering a male bear are greater. Thus, males are more vulnerable to trapping and to "still" hunting, where the hunter waits in one place to encounter a bear. However, if any of these hunting methods are intensified, and if hunting practices such as pursuing bears with dogs are permitted, more hunting pressure tends to focus on the less mobile female bears. Regulating overall hunting pressure, as well as the methods of hunting, can help select for male bears, insuring that fewer females are killed, and thereby helping with the long-term conservation of a hunted bear population.

If food conditions are poor, fewer bears should be killed as they reproduce more slowly under such circumstances. For this reason, it is very important to fine-tune hunting regulations to the biological needs of the local population being hunted. Such regulations must set realistic estimates of the numbers of bears that may be legally removed from the population. As with grizzly bears, allowance must also be made for additional kills through poaching and nuisance-bear removals. Illegal killing of black bears should be minimized by stepping up enforcement of hunting laws, which, in virtually all Canadian jurisdictions, means hiring more conservation officers.

3. *Controlling International Trade in Bear Parts*
In order to place strict controls on trading in bear parts, Canada urgently needs clear legislation, and strong enforcement of that legislation, at the federal, provincial, and territorial levels. Table 4 summarizes the current situation in Canada. If this trade cannot be adequately controlled, it should be banned entirely.

Internationally, it would help if the black bear was placed on Appendix II of the Convention on International Trade in Endangered Species of Wild Flora and Fauna (CITES), especially for Canada, the United States, and Japan (see Appendix D). Although the black bear is not yet endangered by trade, its listing on Appendix II would at least insure that a certificate of origin or an export permit is required for parts shipped from the country of export. This move, however, would permit authorities only to *monitor*, not necessarily to *control* or to *limit* such trade. Also, it would not affect trade within Canada.

A new federal Wild Animal and Plant Protection Act is currently being drafted. This new federal act, incorporating the old Game Export Act, must be used to establish a means of limiting the volume of bear parts not only exported from Canada, but also traded between provinces and territories. The 1990 federal Green Plan promised to introduce this legislation by 1991, but it will require public vigilance to insure that an effective act is produced.

Provincial and territorial governments must decide whether or not they want to permit this trade at all. Each province or territory is responsible for all wildlife parts that are exported or imported into its jurisdiction — a matter that obviously requires cooperation with the federal government as well. However, provinces and territories must also become responsible for controlling and/or banning trade in bear parts within their own jurisdictions, because there appears to be a growing demand for them within Canada. They are readily displayed and available in Asian medicine stores in major cities such as Vancouver and Toronto, and illegal trade in smaller communities may be happening as well.

If passed, all these laws must be backed up by undercover,

intelligence, and forensic efforts to better understand the patterns of trade in bear parts in and out of Canada. These efforts, in turn, will need to be supported by adequate enforcement activities, which successfully apprehend and punish poachers, as well as buyers who operate outside the law. Cooperation between federal agencies such as the Royal Canadian Mounted Police (RCMP) and Canadian Customs officials, and provincial agents such as conservation officers, will be required. There is no reason why hunters' and anglers' groups, or fish and game associations, shouldn't also play a strong role. Voluntary programs to "report a poacher" have already been undertaken by some of these organizations, and they could now be in the forefront, supporting legislation as well as mounting voluntary programs to help shut down an insidious new threat to wildlife populations throughout Canada.

A crucial question regarding all trade in wildlife is whether to allow the trade while attempting to control it, or to ban it outright and run the risk of driving it completely underground. In either case, there will likely continue to be illegal operations as long as the trade remains lucrative. The problem with experimenting with a period of control, then implementing a ban if necessary, is that we run the risk of waiting until it's too late. Right now, in Canada, there are either no controls or, at best, differing laws that have the effect of simply "monitoring" the trade in bear parts. And these relatively weak measures are enforced by a uniformly inadequate number of wildlife officers. The result has been a legal vacuum, which is an invitation to disaster for our temporarily abundant black bear populations.

It would be a great exaggeration to depict the most secure member of the large carnivore family in Canada as "endangered," in the sense of its being on the brink of extinction. However, it would also be a serious mistake to be complacent about the black bear's future. These predators are losing habitat, are heavily hunted every year, and appear to be next in line for significant additional pressure through illegal international trade. Clearly, black bears must be part of a conservation strategy for large carnivores in Canada. As part of our nation's wildlife heritage, they must be maintained.

5.
The WOLF

There were only fourteen wolves in the last wolf pack in Montana.

But it took ten years to hunt the last one down. This talk of reintroducing

wolves just goes to show you that you can send a city boy to college and

teach him wildlife management, but it still don't mean he knows a

damn thing about animals. . . . Those silly jackasses, they'll want

to reintroduce the sabre-tooth tiger next!

LESTER JOHNSON, FORMER RANCHER, NORTHERN MONTANA

LESTER JOHNSON LIVES IN A PART OF THE WORLD WHERE 80,000 WOLVES were poisoned, trapped, or shot under a state bounty system between 1883 and 1918. Up to 1942, another 25,000 were wiped out under a national law to exterminate wolves from all U.S. federal lands. Today, by contrast, the wolf is listed in the U.S. federal Endangered Species Act. And, ironically, wildlife officials in Montana are trying hard to protect one pack in Glacier National Park, which, in the 1980s, moved on its own south from Canada, established a den, and began raising pups.

This snippet of history speaks volumes for what has happened to the wolf. It is true not only for the United States, where the wolf has been eliminated from 95 percent of its range, but also for Europe, where this species now exists in only eight of the twenty-three member states. As a result, conservationists in Norway, Sweden, and other countries are in a similar position to that of their counterparts in Montana — desperately trying to hang on to fewer than a dozen wolves. The question for Canadians is: "What have we learned from the fate of wolves elsewhere in the world, and what steps are we taking to make sure the same story doesn't unfold in this country?"

WOLF FACTS AND FIGURES

The wolf (*Canis lupus*) is also properly called the "gray wolf" or "timber wolf." Many Canadians, particularly in the rural areas of southeastern Canada, call coyotes (*Canis latrans*) "wolves" or "brush wolves." Wolves and coyotes, however, are two distinct species, seldom interbreeding, even when their ranges overlap. This confusion has resulted in farmers in southern Ontario claiming to have "wolves" on their land, when, in fact, they have coyotes or "coydogs," which are crosses between coyotes and dogs.

The two animals can be distinguished fairly easily. The wolf's face is larger and less pointed than the coyote's. Its tracks are much larger, and the stride or distance between them is farther apart, reflecting the wolf's greater overall size. Generally, wolves are found in more remote, timbered habitat, whereas coyotes adapt readily to agricultural areas and even city ravines. As a result, coyotes are now expanding their range in Canada much more than wolves, especially in the south and east, where wolves have disappeared.

Even where they are numerous, wolves are seldom seen. But they can often be heard howling, especially at night, and they will respond to even an amateur imitation by people. Also, wolf "sign" — tracks, scats, and scent marks — are easily noticed once you know what you are looking for. In this way, it is possible to get the "feel" of wolf country, and to contact wolves in a wild setting, even if you haven't actually seen one.

Despite the wolf's natural elusiveness, in preparing this book we heard stories from people who quite literally "bumped into" wolves. Gordon Haber, who has spent more than twenty-six years studying wolves in Denali National Park, Alaska, told us about an incident that occurred in 1967 while he was looking for a particular wolf. "He knew I was in the area, and I knew he was in the area. I was kind of looking around for him and I came up over a little hill. He came up over the opposite side. We came up to the crest right at the same time, and literally looked each other in the eyeballs! We were about five feet apart, and kind of just stared at each other in surprise!" Lu Carbyn, carnivore biologist for the Canadian Wildlife Service (CWS) out of Edmonton, told us he once almost stepped on a big black wolf, which was asleep in the grass in Riding Mountain National Park, Manitoba: "We looked at each other at very close range, both of us totally perplexed and not knowing how we got into the situation! Of course, the wolf left quickly!"

In physical appearance, the wolf is bigger and lankier than a husky sled dog, with longer legs, larger feet, and a narrower chest. Wolves are built for efficient, long-distance travel over frozen lakes and rugged forested land. They can easily cover 75 kilometers (45 mi.) in a day. John Theberge, who studies the wolves of Ontario's Algonquin Provincial Park, notes that our human sense of mobility is not in tune with what a wolf can do. "They may be back and forth across their territory six times easily and have no trouble running across an animal's scent. We don't realize it, because we plod along on snowshoes and get almost nowhere when we're tracking them." Although it is difficult to describe it, once you have seen that characteristic steady wolf gait, it becomes a never-to-be-forgotten trademark of this animal, easily recognized even from a distance.

Wolves often walk in single file. Paul Joslin, currently director of research and education of Wolf Haven International in Washington State, reports an interesting example of this. It happened a number of years ago when he was a student doing wolf research in Wells

Gray Park for the late Douglas Pimlott: "The wolves had come as a pack, down a steep slope, single file, then they fanned out and went up to my snowshoe tracks from the previous day. When they came up to my tracks, they turned around and went back. They got to that hill, and then all walked back up single file, up the very same tracks they had come down. And you know how when you walk down a hill you take bigger steps than when you walk up? They made exactly the same footprints. If you saw only that one set going up the hill, you'd have no idea that six wolves came down and six wolves went up that same hillside in one set of tracks!"

The color of a wolf's coat ranges from white to black and varying shades of cream, gray, and brown. Grizzled gray wolves, with brownish-black backs grading to brownish-white underparts, are most common. White wolves predominate in the Arctic, and black wolves are not unusual in more southern populations. Wolf fur, as well as coyote, is often used as trim on the hoods of cloth or down-filled parkas because it has some ability to shed frost, although Indians and Inuit prefer wolverine fur for this purpose. Wolf fur is not particularly valuable, so commercial trapping has not been a major factor in reducing the overall population. However, for purposes of wolf control, trappers are sometimes given financial incentives to trap more wolves than they normally would. In this case, wolf numbers may be more severely affected.

There are many different subspecies of wolves in Canada, perhaps as many as seventeen. As is the case for the other top predators, we are not just trying to save "the wolf," but the many distinct types of wolves living in different regions of the country. One factor that differentiates these various subspecies is their size. For example, the wolves of Algonquin Park in Ontario are noticeably smaller than those of the Mackenzie Mountains in the Northwest Territories. It's not surprising, therefore, that adult wolf weights in Canada range from 26 to 79 kilograms (57 to 174 lbs.).

These regional differences also account for the fact that wolf behavior and social structure vary significantly across Canada. For example, wolves feeding on migratory caribou in the Northwest Territories don't behave in the same way as those feeding on moose or white-tailed deer in Ontario. Peter Clarkson, a government wildlife biologist in the Northwest Territories, has found that some wolves there move with the caribou herds; others stay in a smaller area and wait for the caribou to pass through. Also, young wolves in the Far North may leave their packs earlier, and more frequently, to start new packs in a new territory.

By far the most important characteristic of the wolf as a species, what sets them apart in the animal world, is its elaborate social organization. Most people know that wolves usually live in packs with a social hierarchy dominated by the "alpha" male and female. In most cases, the pack is an extended family made up of related individuals. All the little things that go on between individuals in this group enable them to function, in Gordon Haber's words, "as a big superpredator, a superorganism, or one predator with many legs." Because the wolf's social structure is so important for the species's survival, we must know when we are affecting it through such activities as direct killing or altering wilderness habitat. It means, for example, that killing individual wolves from a pack can disrupt their finely tuned social hierarchy, which governs pup production in that family group. It means that it may not be possible to reintroduce wolves into a new wilderness area, unless the entire family structure is kept intact. It means that the hunting strategies of wolves may be less effective if the pack, for some reason, is broken up. And it means that just looking at the numbers of wolves in an area could be misleading because, as Haber says, "we tend to give too much emphasis to the wolf's numerical status, at the expense of the much more basic and more important question of the species's social integrity."

This social integrity can also affect the wolf's reproductive capability. Female wolves reach sexual maturity at two years of age, and males probably don't breed until their third year. This aspect of wolf biology is quite different from that of other top predators, such as grizzly bears, which may not mature for eight years. Because female wolves are ready to produce pups at a relatively early age, there is greater potential for reproduction in a wolf population than in a grizzly or other bear population. Also, wolf litters average five pups every year, whereas grizzlies have only one or two cubs at several-year intervals.

There is evidence that, when a wolf pack's social structure is disrupted — for example, if one or both of the alpha animals are killed — or if the pack is disrupted by wolf control programs, more than just the alpha female may breed and have pups. Or, the females that do breed may have larger litters. That may explain why some scientists have observed that the percentage of pups in an "exploited" wolf population is often greater than the percentage of pups in an unexploited or undisturbed population. This type of behavior may serve as an in-built population survival mechanism, in this case insuring that *more* wolves are produced when necessary.

However, other aspects of wolf biology help to *limit* its numbers. For example, female wolves come into heat only once a year, unlike domestic dogs, which are ready to breed more frequently. Also, in the wolf's social structure, usually only the dominant, or alpha, male in an undisturbed wild pack breeds with the alpha female. Therefore, although other adult wolves in the pack may be capable of reproducing, they do not breed because they are not the dominant pair. Females that do breed may have smaller litters when food is in short supply. And, on average, only 40 percent of wolf pups survive to become adult animals. These are all in-built population-control features of wolf packs, and it is important to understand such features when considering the conservation of this top predator. Says François Messier, who has conducted extensive research on wolves in Quebec, "One misconception is that wolves cannot regulate their numbers by themselves. In other words, if they are not artificially controlled, they will grow ad infinitum. That is really a misconception."

These aspects of wolf biology would also explain why there have often been more wolves observed sometime after a wolf-control program is stopped than there were before such a program was started. A pack's social structure may be disrupted by these programs, causing more of the females to breed and to have pups. Or such programs may temporarily reduce wolf numbers, causing a situation where there are few wolves but an abundance of prey, in which case the wolves may have larger litters in response to the improved food supply. Or, since wolves are territorial, if one pack is broken up, then a nearby pack may "immigrate" or move into the other's territory. Or, individual wolves may leave a larger, healthier pack to "disperse" and start a new group in the vacant territory or "niche" created by the pack that was killed off. For all these reasons, wolf numbers can rebound very quickly after a wolf-killing program, often within a couple of years. This somewhat self-defeating effect has been observed many times in connection with wolf-control programs. It means that if such programs are to be "effective" in controlling wolf numbers over the long term, they may have to be continued almost indefinitely.

These very sophisticated forms of natural population regulation have been observed but not well understood by wildlife biologists. Most of us learned in school that predator numbers tend to fluctuate in relation to prey, and some of the mechanisms outlined above help to explain how these changes in wolf populations are brought about.

As well as an intact social structure and adequate territory, to exist on a year-round basis wolves need a food supply of large ungulates such as moose, elk, caribou, deer, mountain sheep, and even bison. Contrary to a popular misconception, wolves do not "just eat mice." However, they may "switch off" to other species such as beaver, especially in summer, and they also eat small animals such as rabbits and various small rodents (including mice), as well as birds, fish, fruit, insects, and even grass. Nevertheless, ungulates must be present for a wolf's long-term survival. These facts have been widely misunderstood, and have led to distorted stereotypes of the wolf. Bruce McLellan, a wildlife biologist from British Columbia, puts it this way: "There's one misconception that wolves are vicious killers — all they want to do is kill people and anything that moves. And the other misconception is that they're pacifists, that they don't kill anything except mice. Let's face it. They eat large ungulates."

The wolf's excellent eyesight is an important factor in its search for food. Gordon Haber has seen wolves stop on peaks and ridges in Alaska, for example, and scan the terrain below to spot moose in forested areas many kilometers away. He has also seen them "picking out white mountain sheep against a snowy background when they're not even moving." Haber is fascinated by their hunting tactics: "They'll see a moose in the distance, and, while they're still several miles off, individuals of a pack will spread apart, trying to anticipate potential escape routes that the prey may take. They do it in a very deliberate way. Often, individuals will separate off, miles apart. You see them sitting there for fifteen or twenty minutes, looking at each other, back and forth over those great distances. And then they finally launch a coordinated maneuver from all these different directions. It's something you can't believe until you have seen it. The best words I have to describe it are 'astounding, phenomenal'!"

Their food requirements, of course, can bring wolves into direct competition with humans who also want to hunt ungulate species. Where human settlement occurs in wild areas, wolves may prey on cattle, sheep, horses, pigs, poultry, and even dogs. As a result, people come to fear or hate wolves, which leads to the kind of widespread persecution and extermination campaigns undertaken in the United States at the turn of the century.

Adult wolves consume between 2.5 and 5.5 kilograms (5.5 and 12 lbs.) of food per day, on average. However, theirs is often a "feast or famine" situation; they may go as long as two to three weeks without food. Although much has been made of whether wolves kill more than they can eat, this debate seems to be rather futile because wolves return to carcasses that are not completely consumed. Such kills also serve as food for many other wildlife species, including ravens, foxes, and wolverines. Therefore, it is difficult to understand any sense in which wolf kills are "wasted." In fact, carcasses are important as a source of food and are efficiently used by a variety of species in natural systems all over the world.

Finally, the relationship between wolves and their prey can be somewhat surprising. Peter Clarkson made these observations in the Northwest Territories: "There is often an air of calmness between caribou and wolves when they're in the same area, before and even after a kill. The caribou definitely know the wolves want to eat, but, until the wolves actually make an advance, the caribou can be quite relaxed, lying there, feeding, and keeping an eye on the wolves. No one seems to be that excited. As the wolves get too close, the caribou take off and run, stop, and, if the wolves pursue, then they keep running. When the wolves do pull down a caribou, then the others stop. They seem to say, 'Okay, it's over for now, no one else is in danger, so let's just resume what we were doing and carry on with life.' In human terms, we would be more cautious, or have more concern about the wolves, whereas the caribou don't seem to at all."

Interestingly, our human caution and concern may say more about ourselves than about wolves. Graham Forbes, who works with wolves in Ontario's Algonquin Provincial Park, is most often asked, "Don't you carry a gun?" Says Forbes, "I am quite surprised by that. I guess there's still the perception that wolves are quite dangerous. It's such a joke in the sense that I've never even thought of needing a gun. Two hundred and fifty to three hundred wolves, and half a million people go through the park every year. Most people never even see a wolf. A few may hear them, but there sure isn't anybody being *attacked* by them!"

François Messier has investigated many wolf kills in Quebec, and he reflects on the wolf's tolerance to such human intrusions: "Let's say you're studying wolves and that one of your major jobs is to fly in and locate wolf packs. Very often they are at kills. You land on a nearby lake and you walk directly to the kill. This is analogous to having a very mean dog and removing the food dish from underneath the nose of that mean dog! Yet, in all those situations, the response of the wolf is to walk or run away. You do your business, and, ten minutes later, when you fly back, they are back on the kill. So, basically, they have high fear of humans and they will walk away, even if you approach one of their kills."

Although 15 to 20 percent of the general public still appears to believe otherwise, there is no documented account of a healthy wild wolf attacking a person in North America. The fictional legend of White Fang, by Jack London, which describes wolves devouring prospectors in the Yukon, is entirely fictitious. The same can be said of "Little Red Riding Hood" and "Peter and the Wolf."

Hank and Val Halliday, of Wolf Awareness Inc. in Ontario, have designed a school program under the motto "Through education, dispelling the myth." This fascinating program not only covers material on basic wolf ecology for junior grades, but also allows the students to participate in wolf research, for example, by raising funds for radio-collars. Through the results of radio-tracking by wolf researchers, they learn about what happens to "their" wolf. Hank and Val's advice to anyone interested in truly understanding wolves is: "Don't whitewash the wolf. People either accept them as predators, or they don't."

Even wolves that have been trapped and then approached by people have been observed to be surprisingly submissive. George Kolenosky, a wildlife manager and bear expert from Ontario, tells an interesting story in this regard. Years ago, he was live-trapping wolves to mark them, then releasing them back into the wild in order to track their movements and better understand their behavior. The leg-hold trap he used had one spring removed so it wouldn't hurt a wolf's leg. A chain from the trap was attached to a rock so that a wolf, once it was caught, wasn't completely confined. It could move around a bit by dragging the rock. George caught a big, male wolf, which had wrapped the chain and rock around an old log hanging out over a creek. In those days, researchers didn't use drugs on these animals, so the only way George could get the wolf out was to pick it up in his arms and unwind the chain by literally going underneath the log, over the top, and underneath again. Before he did this, he tied a shoelace around the wolf's mouth to prevent it from biting him.

"I took this wolf, all sixty-five pounds of it, and started unwinding it. I got so busy doing this that I wasn't paying much attention to the wolf. After it was all unwound and I was carrying it up the bank, the student with me started turning about four shades of white. What had happened was that the shoelace had fallen off the wolf's mouth. It wasn't tied up at all anymore! Here was this wolf's head, literally right near my throat because of the way I was carrying it, like a dog. So that wolf, if it had wanted to, could have just gone 'chomp!' right into my neck. But it didn't."

Gordon Haber told us another fascinating story, about something that occurred while he was watching a wolf den in Denali National Park, Alaska. He noticed that the alpha male — a big, beautiful black wolf — was going up to all the other wolves in the way lead wolves do when they start rounding up the pack to hunt. The wolf was waking them up individually, and there was a lot of friskiness and general activity centered on the lead wolf. Gordon remembers thinking, "That's interesting. They've apparently seen a moose or caribou and they're going to initiate a hunt."

However, as he was looking around to see what it was they were getting ready to go after, suddenly they all started running for Gordon, at top speed, with the large male wolf in the lead! At about a hundred feet, the wolves came to a screeching halt and fanned out in a semi-circle around Gordon. The big alpha male came running toward him, barking and growling. "It was the most hideous, scary thing! Just think of the biggest German shepherd you've known, growling and barking at you, full-bodied, full-throated, a really vicious dog . . . and these were much more vicious than that! I was starting to have doubts, even though I'd been a big champion of how wolves never attack people. I was starting to wonder, 'Well, am I going to be the first person to disprove that?' I figured I'd better start doing something about my safety. So, I stood up, just to make sure they understood I wasn't a moose or a caribou or something, and waved my arms and yelled a little bit. That seemed to do it, because at that point he ran back to the rest of them. They all eventually ran off, back toward the den." According to Gordon, this type of encounter has happened to him since, usually when he was partially hidden in brush and crouching down low. He has always been able to defuse the situation simply by standing up.

The misconception that wolves pose a danger to humans may someday be finally laid to rest, as scientists and the public start to understand and appreciate them for what they are: important parts of the natural scheme of things wild.

As explained in Chapter 1, over the long term there is no doubt that wolf predation benefits wildlife systems by insuring that "the best-adapted animals" in a prey population survive. However, it is often assumed that wolves kill only the old, weak, or diseased animals. It should be understood that wolves are really opportunistic, which means they catch and eat whatever they can. That may include healthy animals that are ambushed or disadvantaged by snow conditions, and it definitely includes younger animals, such as caribou and moose calves and deer fawns. The genetic effect of wolf predation is, therefore, likely to be more long term than simply weeding out this year's old or starving prey. In general, traits that benefit either predator or prey are those that get passed on to succeeding generations, whereas traits that have little or no survival value do not. In this sense, the long-term "survival of the fittest" is insured. Because wolf predation plays a key role in this process, wolves should be conserved so they can perform their very important evolutionary part in natural systems, as well as enhancing the experience of wilderness for all who search it out.

WHAT HAS HAPPENED TO WOLVES?

Wolves were once the most widely distributed mammal in the world, found throughout Europe, Asia (excluding the southeast), and North America (including Mexico). Now they are found in only a small portion of their original range, and usually in greatly reduced numbers.

Worldwide, wolves have been classified by the World Conservation

Union (formerly the IUCN) as a "vulnerable" species. However, this definition of "vulnerable" is different from the COSEWIC definition (see Appendix B). For the World Conservation Union, a "vulnerable" species is "likely to become endangered in the near future if the causal factors of its decline continue to operate." This formal, international classification reflects the fact that, except in Canada, Alaska, and the Soviet Union, wolf populations worldwide are either in steep decline or are already reduced to mere remnants.

Ironically, many countries that once persecuted the wolf are now desperately trying to protect it. Wolf populations in Bhutan, India, Nepal, and Pakistan, for example, are listed in Appendix I of the Convention on International Trade in Endangered Species (CITES) (see Appendix D), so international trade in wolves or their parts from these countries is strictly prohibited, or, at worst, limited and legally monitored. In Sweden, Norway, Italy, India, Israel, and Mexico, wolves are supposed to be totally protected by law. In the lower forty-eight American states, except for Minnesota, wolves are listed as "endangered." That means they are protected by federal law, although wolves that kill livestock are subject to control measures, including being killed.

Today, in Europe, wolves are extinct in many countries of their former range, including Austria, Belgium, Denmark, France, Germany, Hungary, Great Britain, Ireland, Switzerland, and the Netherlands (Table 5 summarizes the current status of wolves in Europe). This bleak situation was brought about by the growth of cities and towns in formerly wild habitat. After the natural area and food available for wolves were greatly reduced, they then preyed on livestock, and were feared and persecuted as a threat to people.

The wolf's status in Eastern Europe, the Soviet Union, and Asia is not as clear (see Table 5). The Soviet Union is reputed to have up to 70,000 wolves, which would rival the Canadian population as the largest remaining in the world. However, wolves are still killed extensively in the Soviet Union because farming operations have expanded much farther north there than in Canada, leading to problems with wolves preying on livestock. As a result, a large number of wolves are killed every year, perhaps as many as a third of the entire population. That could lead to eventual extinction of the species there. Already, wolves are being pushed out of a huge portion of their range in a wide northern band right across the heart of the country.

Table 5: CURRENT STATUS OF WOLVES IN EUROPE AND ASIA

Country	Status	Range Occupied	Cause of Decline
Afghanistan	1,000?, in decline	90%	Unknown
Albania	Unknown	Unknown	Unknown
Arabian peninsula	Fewer than 300, in decline	90%	No protection, persecution
Bhutan	Unknown	Unknown	Unknown (but protected)
Bulgaria	100?, lingering, highly threatened	Unknown	No protection, persecution, habitat destruction
China	Unknown	20%	Persecution, extermination efforts, habitat destruction
Czechoslovakia	100?, lingering, in steep decline, endangered	10%	No protection, persecution, habitat destruction
Egypt (Sinai)	30, highly endangered	90%	No protection, persecution
Europe (Central)	Extinct	Nil	Persecution, habitat destruction
Finland	Fewer than 100, lone wolves and pairs	Less than 10%	Persecution, no protection
Greece	500+, viable but declining	60%	Persecution, habitat destruction
Greenland	50?, lingering, threatened	Unknown	Persecution
Hungary	Extinct	Nil	Unknown
India	1,000-2,000, lone wolves or pairs, endangered	20%	Decreasing prey, persecution, unenforced protection
Iran	More than 1,000, viable	80%	Persecution
Iraq	Unknown	Unknown	Unknown
Israel	100–150, lingering, highly threatened	60%	Habitat destruction, persecution
Italy	250, highly threatened	10%	Persecution, habitat destruction, extermination of prey
Jordan	200?, highly threatened	90%	Persecution, no protection
Lebanon	Fewer than 10, highly endangered	Unknown	No protection, persecution
Mongolia	10,000+, viable, in decline	100%	Active extermination efforts
Nepal	Unknown	Unknown	Unknown
Pakistan	Unknown	Unknown	Unknown
Poland	900, viable	90%	Persecution, habitat destruction
Portugal	150, lingering, highly threatened	20%	Persecution, habitat destruction
Romania	2,000?, declining	20%	No protection, persecution, habitat destruction
Spain	500–1,000, threatened	10%	Persecution, habitat destruction
Sweden/Norway	Fewer than 10, highly endangered	Less than 10%	Persecution
Syria	200-500, lingering low density, highly threatened	10%	No protection, persecution
Turkey	Unknown, viable, in decline	Unknown	No protection, livestock predation
USSR (Asia)	50,000, viable	75%	Control programs everywhere, persecution, habitat destruction
USSR (Europe)	20,000, viable	60%	Control programs everywhere, persecution, habitat destruction
Yugoslavia	2,000, in steep decline	55%	Persecution, habitat destruction

Source: J.R. Ginsberg and D.W. Macdonald, *Foxes, Wolves, Jackals and Dogs: An Action Plan for the Conservation of Canids*, IUCN/SSC Canid Specialist Group and IUCN/SSC Wolf Specialist Group (L.D. Mech, Chair), (Gland, Switzerland, 1990), p. 38-39.

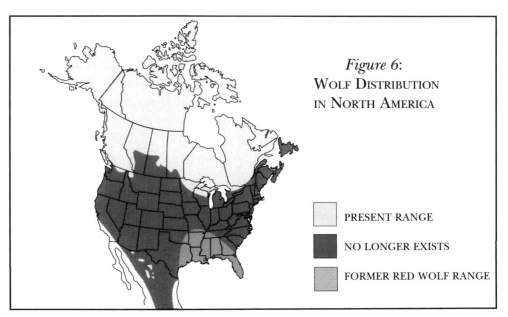

Figure 6:
WOLF DISTRIBUTION
IN NORTH AMERICA

PRESENT RANGE

NO LONGER EXISTS

FORMER RED WOLF RANGE

Source: M. Novak, J.A. Baker, M.E. Obbard and B. Malloch, © 1987, Queen's Printer for Ontario (Toronto: Ontario Ministry of Natural Resources and Ontario Trappers Association), p. 362.

In North America, the pattern has been one of exterminating the wolf relentlessly in the southern part of its former range, so that it now exists only in the northern portion (see Figure 6). Wolves have been exterminated from 95 percent of their former range in the lower United States, and now exist in only four or five states there. They may occupy less than 10 percent of their original range in Mexico, if wolves even exist there at all. Table 6 summarizes the current status of wolves in North America.

Lu Carbyn has an effective way of sharing his perspective on this fact, particularly from his observations in Wyoming, Idaho, and Montana: "When I drove through these huge landscapes and looked at the number of people there, there is still tremendous wilderness left, large tracts of land. It really *is* amazing that humans, over a very short period of time, have been able to completely annihilate the species. I'm talking about huge areas, and pretty light human populations, really."

On a global basis, therefore, Canada and Alaska are likely left with the healthiest remaining wild wolf populations in the world — a unique conservation opportunity and responsibility. Says Montana

Table 6: CURRENT STATUS OF WOLVES IN NORTH AMERICA

Country/State Prov./Territory	Status	Range Occupied	Cause of Decline
Alaska	6,000, viable	100%	Some control programs active
Alberta	4,000, viable	80%	Habitat loss, persecution, agricultural conflicts
Labrador	Unknown, viable	95%	No data available
Manitoba/Saskatchewan	Unknown, viable	70%	Persecution in south, habitat loss, agricultural conflicts
Mexico	Fewer than 10, highly endangered	Less than 10%	Protection not enforced, persecution, habitat destruction
Michigan and Wisconsin	35, highly endangered, lingering	10%	Persecution, habitat destruction
Minnesota	1,200, viable	30%	Persecution, habitat destruction
Newfoundland	Extinct	Nil	No data available
Northwestern United States	30, highly endangered slowly recolonizing	5%	Persecution, habitat destruction
Northwest Territories	5,000 – 15,000, viable	100%	Stable
Ontario/Quebec	10,000? viable	80%	Persecution, habitat loss, agricultural conflicts
Southwestern U.S.	Extinct	Nil	Persecution, habitat destruction
Yukon/B.C.	8,000, viable	80%	Habitat loss, persecution, agricultural conflicts

Source: J.R. Ginsberg and D.W. Macdonald, *Foxes, Wolves, Jackals and Dogs: An Action Plan for the Conservation of Canids*, IUCN/SSC Canid Specialist Group and IUCN/SSC Wolf Specialist Group (L.D. Mech, Chair), (Gland, Switzerland, 1990), p. 37-38.

wolf researcher John Weaver, "For us down here in the United States, it's a matter of basically trying to restore the wolves in a few remnant areas, whereas you folks in Canada are in a much better position at this point. But I find it very sobering to consider that some of the first species to go extinct were the most abundant ones. So, you can't have any false comfort at this point simply because you have widespread populations."

Throughout Canada, before European settlement, wolves were the most widespread mammal. They were found virtually everywhere, except on the Queen Charlotte Islands and Prince Edward Island. After the arrival of European settlers, human pressure from the south on wolves grew because of the expanding human population. Finally, the wolf was exterminated from the Atlantic provinces by the 1900s. It became extinct in Nova Scotia and New Brunswick by 1870, and in Newfoundland by 1911. Wolves have been pushed back from the southern portions of both Quebec and Ontario by the growth of agriculture and cities. Prairie "buffalo wolves" used to prey on the bison, but when the bison were exterminated by 1900, the wolves went with them. Farming and urbanization have also pushed this predator out of parts of southern British Columbia.

Over the past 100 years or so, money has been paid out to kill wolves through bounty systems in Quebec, Ontario, Manitoba, Saskatchewan, Alberta, British Columbia, and the Northwest Territories. Bounties, however, have rarely been proven to accomplish their goal of controlling wolf numbers in Canada. In effect, they have really been an outlet for anti-wolf attitudes more than anything else. Since social attitudes toward wolves have changed, and wolf bounties in this country have not been effective anyway, they have been formally discontinued at the provincial and territorial levels throughout Canada. Nevertheless, some local governments still want to maintain their own bounties. And some local trappers are still paid by governments, or by sport hunters' associations, to kill wolves. This is not a general public legal bounty as such, but it obviously *is* an example of using financial incentives to kill wolves.

When all is said and done, wolves currently occupy about 85 percent of their former range in Canada (see Figure 6). Their current status is summarized by province and territory in Table 6. Although numbers of wolves appear to have increased in some specific areas, these trends follow long periods of large-scale wolf control, such as the poisoning programs of the 1950s and 1960s. Wolves have also made efforts to recolonize some of their former range, for example, in Alberta along the eastern slopes of the Rockies, and into Montana from British Columbia.

Dennis Voigt, from the Ontario Ministry of Natural Resources, warns about being complacent in Canada about wolves because there still are relatively large populations: "Complacency may be an even bigger threat than habitat destruction. Wolves are not in the endangered-species situation that they are in places elsewhere in the world or in North America, so, among many northern people, there's a sense of complacency and they don't relate to what has happened elsewhere."

The overall population picture is one of a relatively stable Canadian wolf population, numbering 50,000 to 65,000 animals. This situation presents Canada with a chance to do something no other country has done: deliberately to conserve healthy wild populations of different types of wolves on one of the last landscapes still capable of supporting such a conservation goal.

CURRENT THREATS TO WOLVES

Simply put, the biggest threat to wolves is people who are not prepared to share food and wilderness territory with them. We establish our farms in wolf habitat, then, when the wolf does what
comes naturally by eating our livestock, our answer is to get rid of
the wolf. Since the wolf hunts the same animals as human hunters,
again our answer is to get rid of the wolf. In effect, by killing the wolf
we're trying to get rid of the competition. And as long as wolves are
seen as competitors, as long as we are unprepared to share with
them, there will be pressure to reduce wolf numbers. Whether they
take the form of wolf-control programs or habitat disturbance,
these activities against the wolf are all secondary spinoffs of basic
human intolerance.

This competitive attitude, reflected in demands for wolf
control, has been heeded by governments in Canada, which have
initiated programs to kill wolves in specific areas where the animal
is judged to be a problem. As long as these programs are very
local, and targeted at discouraging or removing one or two individual wolves that are causing problems for farmers who are
already taking steps to avoid such problems, then the impact on
overall wolf numbers is not likely to be very serious. However,
when wolf-control programs are intended to kill a large percentage, typically 60 to 80 percent of the wolf population in an entire
region, then the impact can be more serious. In northeastern
British Columbia, for example, a wolf-control program resulted in

killing 1,000 wolves over five years in a specific area.

It is important to note that the goal of such wolf-control programs in Canada is not intended to eliminate the wolf entirely. Nevertheless, wolf population estimates are often sketchy at best, so the impact of control programs may be worse than intended if there were, in fact, fewer wolves than originally thought. If such programs were continued over the long term (they may have to be in order to accomplish their goal), and if they were initiated in a number of different areas, clearly the same kind of unacceptable situation could result for the wolf in Canada as now exists in Europe and the United States. Therefore, local control programs must not, in effect, become extermination programs, which result in overkilling at the local level first, then spread to wipe out entire populations. This problem is not restricted to Europe and the United States; it has already happened in the southern portion of Canada.

World Wildlife Fund believes that the only safe conservation route is to stop large-scale wolf-killing programs altogether. As we have already seen, the wolf's reproductive biology is such that control programs may be futile unless they are continued indefinitely. Wildlife managers are therefore confronted with the prospect of being "hooked on management." But, if such programs *are* continued indefinitely, they will likely have a negative long-term impact on wolf numbers. Therefore, such programs appear to be no-win propositions. Furthermore, large-scale wolf-control programs are based on a number of assumptions, all of which are the subject of intense scientific and ethical debate. Examples of these assumptions are that:

- wolves are the primary limiting factor on a particular ungulate population;
- the "management" (protection or increase) of the wild ungulate population preyed upon by wolves requires the "management" (decrease) of the predators;
- reducing wolf numbers to a predetermined level will result in predictable increases in ungulate populations;
- if no wolf control is undertaken, predator and prey systems will stabilize at moderate to low levels, and that this is an "undesirable" situation.

Legitimate questions that have been raised regarding these assumptions are:

- Why do humans want higher numbers of ungulates in the first place, and what will be the impacts on the natural ecosystem if their numbers are artificially increased?
- How important are factors other than wolves in limiting the ungulate population, for example, weather conditions, food supply, human hunting, or industrial activities? Have adequate steps been taken to address these other factors where possible?
- If these other factors have not been addressed, then isn't it possible that reducing wolf numbers may not have the desired effect of increasing ungulate numbers?
- Are wolves being killed simply because humans can't or won't do anything about the other things that are limiting ungulate numbers?
- How predictable are the results of killing wolves? For example, what happens if a severe winter or a disease intervenes to reduce ungulate numbers whether or not wolves are preying on them?
- Is the apparent decline in ungulate numbers really a long-term trend that should concern us, or is it simply part of normal population fluctuations? In other words, is the wolf-control program based simply on a "snapshot" in time, or a longer-term perspective on wolf-prey dynamics?

Some of these are ethical questions — difficult, but answerable based on personal values and political judgments. Others are unanswerable because we don't have adequate field information, particularly for the "multi-predator/prey systems." In this sense, another threat to the long-term survival of wolves is the lack of good biological information, and our tendency to make decisions assuming we know things we don't.

The lack of legally protected areas where wolves can establish adequate territories and interact with their prey under natural conditions, undisturbed by people, is yet another concern. John Theberge estimates that, of all the national and provincial parks, wilderness areas, and wildlife reserves in Canada, only twelve areas, representing 1.2 percent of the current wolf range in Canada, are completely protected. He has found that these areas probably protect about 1,600 wolves, or about 2.7 percent of the total wolf population in this country. Among those, very few packs would have their total territory within a protected area. This consideration is important because, when wolves move outside these protected areas, many are killed by trapping, hunting, or vehicles. Says Graham Forbes, "If there's trapping right up to the edge of a park border, you're essentially trapping a park wolf population." Therefore, if wolves are to survive over the long term, there must be sufficiently large, adequately protected natural areas available to them.

Other threats to wolves that can be important on a regional or more local level are the effects of particularly heavy trapping and hunting (especially when there are no limits on the number of animals that can be legally killed), use of poison as a control measure (especially when it is undertaken illegally by private landowners), and killing of wolves by vehicles on roads, snowmobiles, and trains on railway right-of-ways through wilderness backcountry. John Theberge estimates that, in one way or another, humans kill about 14 percent of the wolf population, or about 8,000 wolves in Canada, every year.

Accessing wilderness areas can be as hard on wolves as it is on bears. Barbara Scott, who conducted wolf research on Vancouver Island, reports that, "once there's a road, it makes the wolves really vulnerable. One pack was in an area that had been logged heavily, creating great visibility and great access. Come hunting season, those wolves were wiped out." Graham Forbes, studying wolves in Ontario's Algonquin Park, also expresses concern about this: "During the summer, we observed someone driving on Highway 60, through the core of the park, swerve to hit a wolf, killing it. The driver got out of the vehicle and bragged about it. We've lost a lot of radio-collared wolves; it's hard to get used to."

A different threat arises from the shadowy question about whether wild wolves interbreed with domestic or stray dogs, and with coyotes in some regions of Canada. This dilution of the wolf gene pool does not appear to be a problem in Canada. But it is worth noting because, in other countries, particularly Italy (which has nearly 300,000 feral and wild dogs), it has become a major conservation concern.

Another hybridization concern arises as a result of people deliberately crossing wolves with dogs, and selling the pups as some kind of "superpet." Paul Joslin, of Wolf Haven International, a captive-wolf facility, says that, "in the case of hybrids, it's a very broad problem across the Continent. Clearly, there are a lot of people out there with wolf/dog hybrids. And most of them get into it inno-cently enough. The animal as a pup looks very enticing, and they take these animals on, only to discover that they don't turn into nice, big watchdogs. They're not at all suitable as pets." This concern was recently addressed by the IUCN Wolf Specialist Group, which, in 1990, drafted a special resolution on the problem of wolf/dog hybrids (see Appendix G).

BLUEPRINT FOR SURVIVAL

The conservation measures outlined in Chapter 8, which cover all large carnivores in Canada, are important to the future of the wolf as well. In addition, however, a number of specific steps apply to wolves in particular.

1. *Ethical Questions*

The ethical question of a wild animal's right to exist, and the extent to which we are prepared to share wild prey with other predatory species, undoubtedly applies more to the wolf than to any other large carnivore. These questions were posed many years ago by the late Douglas Pimlott, first chairman of the IUCN Wolf Specialist Group and member of the WWF Role of Honour. Such concerns will probably never be addressed consistently across such diverse interest groups as sport hunters, aboriginal peoples, guide outfitters, government officials, academics, and wildlife preservationists.

The IUCN Wolf Manifesto (see Appendix A) asserts that "wolves have a right to exist in a wild state . . . in no way related to their known value to mankind . . . with man as part of the natural ecosystems." The manifesto also makes it clear that the wolf has been denied this right as a result of a "harsh judgment" based on fear and hatred and, further, that this fear has been "based on myth rather than fact." However, the Wolf Manifesto also notes that "there has been a marked change in public attitudes towards the wolf," causing governments to "revise and even eliminate archaic laws." Although the Wolf Manifesto was drafted in 1973, the trends it identified have only intensified since then.

Clearly, the majority of Canadians today identify wolves in a positive way with wilderness, and many people want wolves to be conserved, not driven from the Canadian wilderness scene. In the terms of Alvin Toffler, author of *Future Shock*, this "new wave" of thinking is now colliding and rippling through the "old wave" of thinking. The result is "choppy water," which will require some patience and rethinking by those advocating one or the other viewpoint.

The new wave, the wolf-protection/conservation advocates, will have to avoid exaggeration. Wolves are not an endangered species in Canada. In fact, their numbers are still relatively high and stable. Adherents to this school of thought will also have to realize that wolves do cause problems for some livestock owners, and that these problems deserve respectful consideration, which may include removing or killing individual wolves when they are clearly identified as problem animals, and paying compensation in the form of cash to farmers for livestock lost. Finally, provided that hunting and trapping are carried out within safe conservation limits, these uses should be respected by wolf protectionists when such activities are for subsistence purposes.

The old-wave thinkers — the Lester Johnsons of this world — must realize that times and attitudes have changed. Many animals, including the wolf, have become genuinely endangered elsewhere because of outdated human attitudes and practices such as extermination programs. It is not just a matter of do-good "city people" naively thrusting hardship on those who live in the country or the bush. Wildlife is now being perceived as having value, not just as something to simply get rid of, or to be hunted.

No one is unprepared to address legitimate problems caused by wolves. But those who have problems must also indicate clearly that they have taken every reasonable step to avoid such problems in the first place. Thus, farmers should properly fence and protect livestock, and dispose of carcasses in such a way that they don't attract predators, which then become a "problem." Also, hunters must do their share in conserving species they want to hunt by accepting reduced bag limits on deer, moose, and other game animals, rather than requiring the wolf to hold back. And hunting must be accepted as an activity that takes place within a natural system that produces both lean years and good years in terms of the supply of game. Since wildlife populations fluctuate under natural conditions, not every year can or should be a "good year." To manipulate the competitor — in this case, the wolf — in an effort to insure that every year is a good hunting year is to behave most unnaturally and unlike any other natural predator. This is not an argument against hunting; it is an argument for ethical hunting.

2. *Controlling Wolf Control*

If the foregoing arguments were accepted, there would be little need for wolf control in Canada. In fact, it is very likely that, some day, Canadians will look back on wolf-control programs and see that they have outlived their usefulness, scientific justification, and social acceptance. Therefore, to the extent that such control programs continue, they will remain controversial.

Of course, wolf-control programs *should* be controversial. They warrant public debate, given the disastrous effect they have had when conducted on a large-scale, sustained basis on wolf populations elsewhere in the world. As long as they continue to be proposed, it will be important that wolf-control programs continue to be subjected to public and professional scrutiny. British Columbia, for example, is experimenting with a Wolf Working Group, which involves diverse public-interest groups in provincial wolf-management policy. Although it is too early to judge its success, this forum deserves to be supported in principle, as does independent scientific review of proposed wolf-control programs. Other jurisdictions in Canada must also bring public input into their wolf-management policies and practices.

Wolf-control programs aren't always large in scale. Sometimes they involve individual landowners trying to cope with wolves killing livestock. One very promising way for farmers and ranchers to deal with a few specific problem wolves, without having to kill them, is by using livestock guardian dogs.

Livestock guardian dogs are an ancient European predator-control technique that is finding new life in North America. These are not attack or herding dogs. They have been selectively bred for hundreds of years to protect livestock. There are currently seven specific breeds of livestock guardian dogs that are used most often. These dogs are large, usually weighing 35 to 45 kilograms (78 to 100 lbs.), and stand 65 centimeters (26 in.) or more at the shoulder. They have medium to long hair; drooping, hair-covered ears; relatively short muzzles; and long tails. In a way, they frequently resemble sheep, the livestock they have guarded most often in the past!

The key to the success of a livestock guardian dog is its attitude. It is trustworthy, attentive, and protective. These traits are instinctive, but proper handling and training are required to hone them for the particular livestock they are to guard. Trustworthiness prevents the livestock guardian dog from being a threat to the livestock it guards. Attentiveness keeps it with the livestock at all times, an especially important trait because, if the dog were to chase after a coyote, wolf, or cougar, the livestock would be left unprotected and vulnerable to other attacks. Finally, protectiveness is the trait that enables the livestock guardian dog to convince potential attackers to leave, through "attack and retreat" behavior by the dog.

Studies have been undertaken in the United States to test the effectiveness of livestock guardian dogs in reducing predation. In most cases, the results are excellent. For example, various research projects have shown that up to 75 percent of the farmers and ranchers who used livestock guardian dogs were satisfied with the results. In an Oregon State University study, two-thirds of the farmers and ranchers stated that livestock guardian dogs reduced the incidence of livestock predation substantially, while one-third said use of the dogs completely eliminated the problem. Sixty percent felt they relied less and less on other forms of predator control after using these dogs, and 50 percent said they no longer had sheep confined at night.

In provinces and territories where there are no livestock compensation programs, perhaps governments could be convinced to subsidize farmers in acquiring their first livestock guardian dogs. Even where compensation programs exist, governments should be encouraged to assist farmers who want to purchase these dogs. Reactive financial costs exchanged for proactive action without killing predators is surely a worthwhile goal. Further information on livestock guardian dogs can be obtained by referring to the "Contacts" section of this book.

3. *Protected Areas*

The movements and fate of wolf packs within and outside protected areas are of increasing conservation concern. Recent studies in Alberta's Jasper and Banff national parks, Riding Mountain National Park in Manitoba, and Algonquin Provincial Park in Ontario indicate that wolves routinely "spill" outside such protected areas. When they do, they are often killed. As a result, Algonquin Park wolves, once considered "protected," should now be regarded as an "exploited" population. This evidence reinforces WWF's recommendation that no predator control should be permitted within protected areas such as national and provincial parks and ecological reserves.

4. *Hunting Policy*

Where wolves are listed as game animals, they should be actively managed as such, with data collected to indicate population levels, and regulations that include specific areas where hunting seasons and bag limits are in effect. Wolves should not be simply huntable without bag limits, or without seasons, or without area-specific regulations. Such is a recipe for overexploitation, which says that the wolf is held to be of little worth by the government agencies responsible for managing wildlife and enforcing hunting regulations.

5. *Recolonization and Wolf Recovery*

Field evidence indicates that wolves are attempting to recolonize areas such as the southern portion of the eastern slopes of the Canadian Rockies, and southern British Columbia into northern Montana. It is not clear whether wolves can be successfully reintroduced into an area by people. Consequently, natural immigration or recolonization of former range may be the only way to re-establish wolf populations in former wolf range. Because such recolonization is a positive development, for conservation purposes and for ecosystems as a whole, it should be encouraged by easing or closing hunting and trapping seasons in areas where recolonization is taking place, *and* where recolonizing wolves originate. Recolonization should not be discouraged by maintaining these pressures, or by initiating wolf-control programs because the animal is beginning to reappear. As John

Weaver says, "This whole idea of restoring a large and controversial carnivore is a major step forward in practicing a conservation land ethic."

One of the biggest blocks to the restoration of wolves in areas where they have been extirpated is "people politics." Lu Carbyn states the problem clearly: "To bring the wolf back is a very special challenge. There's a lot of human selfishness involved. Selfishness on the part of those who want the wolf back, who sit in their city and urban environments and want it back in somebody else's backyard, where it may become a problem for those people living there. And selfishness on the part of those who don't want the wolf back, who say, 'We don't want a competitor here; we want this for ourselves!' The selfishness goes through both camps."

6. *Wolf Appreciation*

So much of the writing about wolves deals strictly with their biology, or states the seriousness of their worldwide situation, that we are in danger of ignoring the more joyful aspects of their lives. Dick Dekker, who has spent more than twenty-five years observing wolves in Jasper National Park, Alberta, had this thought one day while watching wolves on the flats, feeding on the carcass of a big elk: "I sat there with my binoculars and telescope, watching the wolves. It struck me that anyone who doesn't like wolves should have the chance to observe a pack in the wild. There is, in my opinion, nothing as happy, boisterous, and fun to watch as a pack of wolves that includes four or five pups having a whale of a time! There was no fighting, no snarling. That wild pack was relaxed, on its territory, had plenty of food, and they were the most playful animals you've ever seen."

Perhaps more than any other experience, the howl of a wolf is associated with wilderness. This widespread feeling, coupled with the fact that wolves can be readily encouraged to respond to human howling, presents Canadian wildlife biologists and interpreters with an exceptional opportunity to encourage public understanding, appreciation, and support for conservation of this species.

Paul Joslin, who helped pioneer wolf howling as a scientific technique to locate wolves in Ontario's Algonquin Provincial Park, remembers: "The work on the individual packs revealed that, yes, the wolves will consistently respond throughout the summer. After a while you can get to know who's who, who's home tonight and who's not, on the basis of different pitches in howls, whether they're yippy-types or long, drawn-out . . . whatever. Also, we began to unravel characteristics associated with howling. We tried everything, from comparing the moon versus no moon, warm weather versus cold weather, etc., and the one thing that was linked with howling was the wind. If there was much wind at all, the wolves would not respond."

From 1963 to 1980, over 38,000 people attended public wolf howls in Algonquin Park. Responses were obtained on 55 percent of the attempts. This type of program could be initiated by other naturalists in Canada, but, of course, they must be undertaken in a controlled, respectful manner so that they in turn don't have a negative impact on wildlife or wilderness habitats.

7. *Canadian Strategy*

Wolves are the most rapidly reproducing of the large carnivores, and Canada probably has the world's healthiest remaining wild population. A national strategy for wolf conservation, therefore, should be directed toward making sure that this situation does not erode, as it has in so much of the wolf's range elsewhere in the world. Canadian wolf-conservation goals should specify not only the overall numbers that must be maintained, but also that the different subspecies and types of wolves which have evolved in the diverse regions of our country must be protected. This principle is supported by the IUCN Wolf Manifesto.

Since wolves don't establish territories that align neatly with human political boundaries, the provinces and territories of Canada will have to cooperate and compare results of their respective conservation efforts if these important national wolf-conservation goals are to be accomplished.

8. *International Cooperation*

Canada should continue its full participation in the IUCN Wolf Specialist Group. Such participation accomplishes three things. First, it sensitizes Canadian experts to the concern of the international community regarding the fate of the wolf in our country. Second, it provides an opportunity for the Canadian public to access information regarding the status of the wolf throughout the world. Third, it provides Canadians with an opportunity to tell the rest of the world what we are doing to conserve the species. With luck, hard work, and some vision, we will have a good story to tell.

It is easy to make a case for saving endangered species whose "backs are to the wall" because their numbers have dropped below some critical point. But how much easier it would be if we could head off these crisis situations by being more farsighted, by taking steps to conserve wildlife species *before* they reach the crisis point. This is precisely the challenge posed to Canadians by wolves — still regarded by some as vermin; still the subject of unfounded fear; yet, still with us in sufficient numbers that conservation measures can make a difference.

Tragic mistakes have been made elsewhere with regard to this species. The sorry record of extermination is there, clearly written, for all to consider. There is no excuse for presiding over the loss of the wolf from Canada. We even know what to do to prevent that loss. So, the big question remains: "Will we do it?"

6.
The COUGAR

After all the stomping around I've done in the bush, I'd say one of the

highlights in my life has been encountering cougars. On one occasion,

I crept into a meadow. It was autumn; tall grasses were coming up

out of the snow. And, there was this family of mountain lions, about

fifty meters in front of me. I watched them playing in the evening sun

for over an hour. It was just fabulous! There were two kittens and a mother,

but they were all almost the same size. The kittens were fighting like house

cats. It's surprising to me that more people aren't interested in the cougar.

BRUCE McLELLAN, WILDLIFE BIOLOGIST, BRITISH COLUMBIA

MOST PEOPLE ARE AWARE THAT THERE IS SERIOUS CONSERVATION CONCERN about the world's big wild cats, especially spotted cats such as the leopard, cheetah, and jaguar, as well as the spectacular African lion and Indian tiger. However, few North Americans know very much about our own large wild cat — the cougar, also known as the "mountain lion" or "puma." This lack of knowledge is probably related to the fact that few people actually see these animals in the wild. In fact, even people who live near prime cougar country are surprised to hear that they are there, and when they do learn of it, they assume the cougar must be endangered because it is so seldom seen.

There are actually two distinct populations of the cougar in Canada — one in the East, which is very endangered, the other in the West, which is not. What has happened to the eastern cougar should serve as an instructive warning to us all: we must make sure that the same thing never happens to the western cougar.

COUGAR FACTS AND FIGURES

Canada's cougars really *are* large carnivores. In general size, they fall between bears and wolves. A big male cougar can weigh up to 125 kilograms (278 lbs.) and measure 2.5 meters (8 ft.) long. However, most males average 60 to 70 kilograms (133 to 156 lbs.), and females weigh about 45 kilograms (100 lbs.). This makes cougars about the same size and weight as the average adult human.

These beautiful wild cats are a tawny to grayish brown, with a noticeably lighter buff-white belly and throat. Usually, the back of the cat's ears, tip of the tail, and stripes on the muzzle are black. A cougar's coat is short, and not used as a commercial fur. The summer fur is lighter and brighter than the darker, longer winter coat.

The cougar is a long, lithe, powerful animal with big paws and a long cylindrical tail. In deep snow, its tail, which measures about 75 centimeters (30 in.), may show up as a drag mark. Relative to the rest of its body, the head is quite small, with short, rounded ears. Nevertheless, the cougar's skull is wide and strong, its jaws short and deep, and its canine teeth or "fangs" well developed for the challenging job of catching deer and eating meat. Its front paws have five claws, the hind paws four. These claws seldom show up in the tracks or paw prints. The cougar has a habit of putting its hind foot in exactly the same track as the forefoot when walking, a trait typical of an animal that uses stalking as a hunting technique. Cougars are excellent stalkers and climbers, and they can swim when necessary.

Three subspecies of the North American cougar (*Felis concolor*) are found in western Canada: the Vancouver Island cougar (*Felis concolor vancouverensis*) is found on Vancouver Island and on some adjacent Gulf Islands (Saltspring and Quadra); the coastal cougar (*Felis concolor oregonensis*) is found in the coastal mountains and slopes of British Columbia north to the Bella Coola Valley; and the mainland or Rocky Mountain cougar (*Felis concolor missoulensis*) is found in the rest of British Columbia and western Alberta. In designing conservation strategies for western cougar populations, we must be sure to maintain healthy populations of each of these three subspecies.

The eastern cougar (*Felis concolor couguar*) once occurred in New Brunswick, Nova Scotia, Quebec, Ontario, and perhaps Manitoba. Exact differences in appearance between western and eastern cougars are not clear, but the eastern subspecies may be somewhat smaller and perhaps more reddish, with darker coloration along its back. This subspecies of cougar is definitely endangered, perhaps even extinct, although various sightings are still reported from its former range in eastern Canada.

Cougars have been variously described as wary, noiseless, secretive, and elusive. They are active at night, but most active during dawn/dusk periods, which accounts for why they are not often seen. Like bears, cougars are solitary animals, except when they are mating or when females are traveling with kittens.

Unlike other large carnivores, and most other wildlife species, cougars do not breed during one particular season of the year. Female cougars can give birth throughout the year, though more kittens are born in the summer than in any other season. Females reach sexual maturity at about two and a half years of age, and some may not breed until about age four. Their litter sizes range from one to six kittens, and the average interval between births is about one and a half to two years.

Cougar kittens are born with blue eyes, which gradually assume their adult yellow color. Their light, buff-colored coats have dark spots and bars, which generally fade by the time they are nine to twelve months old. Kittens or juvenile cougars stay with their mother until they are one to one and a half years of age, and are dependent on the adult cat until that time.

Females appear to begin their cycle and breed again soon after their current litter either leaves or is lost. Therefore, if the average wild female cougar lived to the age of twelve years, in her lifetime she would likely produce eight to twelve kittens, only two-thirds of which may survive to independence. Thus, cougars are much closer to bears than to wolves in terms of their reproductive capability. When considering the conservation of cougars to maintain viable long-term populations, care must be taken to protect females, especially females traveling with kittens, and to make sure that we establish measures that respect the cougar's relatively slow reproductive rate.

Cougars throughout Canada feed primarily on deer species when they're available, especially white-tailed deer, mule deer, and Vancouver Island's black-tailed deer. However, they will also eat elk, moose, big-horned sheep, mountain goats, porcupines, beaver, snowshoe hares, ground squirrels, mice, birds, and on occasion even coyotes, foxes, lynx, bobcats, and skunks. Unlike bears, cougars prefer fresh meat, and do not usually feed on carrion or garbage. If food is scarce, or if a female is feeding kittens, they will sometimes cover fresh-killed prey with grass, sticks, and leaves, then return to feed on it again. Captive cougars are known to eat an average of 4 kilograms (9 lbs.) of meat per day, but no reliable estimate for wild cougars has been made so far.

When cougars hunt, they usually stalk their prey, springing from bushes, rocky cliffs, or other hiding places, onto a deer's shoulder or back when they get close enough. They don't engage in long chases, but, as Ian McTaggart-Cowan says, "They try to get uphill and then make several tremendous bounds downhill." The force of the cougar hitting its prey is believed to sometimes kill the deer quickly, but most often, with deft paws while astride the deer's back, the cougar breaks the deer's neck.

Cougars will eat most species of livestock, although this problem appears to be much more prevalent in the southwestern United States than in Canada, where they don't seem to bother free-ranging cattle to any great extent. Martin Jalkotzy, a Canadian cougar biologist in Alberta, reports that very few ranchers he knows consider cougars a pest or a serious threat to their livestock. For years, there have been only four or five incidents of reported livestock losses per year along the eastern slopes of the Rocky Mountains, where this problem would most likely occur. Daryll Hebert, a government wildlife biologist in British Columbia, reports that 10 percent or less of all complaints regarding predation by wild animals on livestock in that province involve cougars. It appears unlikely that large-scale programs to control problem cougars preying on livestock will, or should be, a conservation threat to this predator in Canada. However, if current trends pushing agriculture into wild upland areas continue, this situation could change.

The cougar's food habits offer a few points of conservation significance. First, cougars are so closely tied to deer species as their main prey that conserving cougars often amounts to conserving deer populations and habitats. Thus, regulations on how many deer are killed through human hunting should be fine-tuned to insure that enough deer are left for cougars, rather than killing cougars to make sure enough deer are left for people. Second, cougars require habitat with some variability in topography, such as trees, cliffs, draws, bluffs, shrubs, and other features, which they can use for successful stalking. Wide-open, tree-cleared rangelands, farmsteads, or residential developments do not make good cougar country.

Cougars have been known on rare occasions to attack people, especially small children. One of these attacks may have been attributable to rabies, some may be deliberate predation by starving individual cougars, and some may be examples of female cougars attempting to defend their kittens, especially when litters are very young. Ian Ross, a cougar researcher from Alberta, describes an interesting incident that occurred while he was walking in on a radio-collared female cougar. When he first saw her, she was only about 25 meters (27 yds.) away, and she started to moan — a sound that Ian interpreted as a threat. He started to speak to her, and she turned her head slowly and looked at him, continuing to moan. Ian was prepared simply to leave through the canyon, but the cougar was standing in the only exit. She took three or four steps, very slow steps, toward him, and he saw her gaze become very focused — that intent gaze that a cat produces when it spots a bird out in the backyard. "Then she broke into two or three very fast steps toward me, and I thought, 'This is it!' I just shrieked at her and waved my arms, and she stopped. She turned, and then very slowly walked out of my way, moaning constantly. I couldn't keep continuous eye contact with her because the cover was so dense, but I knew where she was and she got out of the way. So did I. I cleared out of there!"

A couple of days later, when the cougar was away, Ian went back and found two newborn kittens. They would have been about seven days old when he and the cat had had their first encounter. "They were just tiny little things, the size of plump red squirrels. If ever an animal has justification for defense, that's it. I was within a few meters of those kittens, so I was asking for it. I don't fault her at all. In fact, I would have felt badly if she wouldn't defend newborns like that."

Martin Jalkotzy speculates that some of what is perceived to be cougar aggression toward people may actually be curiosity or even playfulness. While doing a summary for the Canadian Parks Service, he documented a case where a cougar ran after a young girl and walked up and touched her with its nose, but didn't do anything else. In another case, a photographer was trying to get a picture of a cougar down in a ditch. The cougar started approaching him, so the photographer ran away, across a parking lot. The cougar chased after him, and he managed to fend it off with his tripod. The only injuries sustained were by the photographer, on himself, with his penknife! These, and other cases where cougars have bounded after people running away, have caused Martin to conclude: "I suspect that a lot of what we perceive as threatening is really curiosity — watching the humans run. 'Hey, look at this human run! Not bad, I'm almost loping'!" Martin felt that, in all these cases, if the cougar was really interested in predation, it would have attacked and done real damage.

Nevertheless, there are documented cougar attacks on people in Canada and the United States, some of which have resulted in fatalities. In British Columbia, there were fifteen verified cases of cougars attacking people up to 1976, and nine additional attacks and two deaths up to 1990. Jay Tischendorf, a cougar researcher from the United States, tries to put these, and attacks by other top predators, into a broader perspective: "It is obviously not true that carnivores across the board are aggressive or savage toward humans. Rather, humans have a way of getting themselves into bad situations with any kind of wildlife, from rattlesnakes to chipmunks and squirrels. I think probably every animal in the world at one point or another has 'attacked humans,' but probably it almost always relates to the person being in the wrong place at the wrong time. And, not always, but in many cases, the human was harassing the animal."

WHAT HAS HAPPENED TO COUGARS?

The cougar's range was once the largest of any land mammal in the Western Hemisphere, ranging from the tip of Chile in the south to the southern Yukon in the north, and from the Atlantic to the Pacific across North America. The historic Canada/United States range was about 8.9 million square kilometers (3.4 million sq. mi.). This has been reduced by over 50 percent, to about 3.9 million square kilometers (1.5 million sq. mi.) today (see Figure 7).

Although they have not been trapped for commercial uses as "furbearers," cougars have been persecuted as part of a larger campaign to eliminate large predators overall, through policies such as bounty systems and cougar-control programs to protect livestock and to provide more wild deer for human hunting. More important, the spread of agriculture from east to west converted most of the cougar's wildland to a cleared or settled human-dominated landscape, which violated this cat's need for proper habitat, adequate prey, and solitude.

As a result, in the United States, viable numbers of cougars are now found only in the West. Although the cougar is not "abundant" anywhere, most are found in Idaho, Utah, Arizona, California, and some parts of Montana, Colorado, and New Mexico. Smaller populations are still found in Nevada, Wyoming, Oregon, and Washington. Very small populations may still exist in western Arkansas and eastern Oklahoma. And a remnant, endangered population of the eastern cougar subpopulation is protected on the southern tip of Florida.

In Canada, the most frequently asked question with respect to the eastern cougar is, "Does it still exist?" There is no doubt that cougars once existed in eastern Canada, as part of a larger population extending up from the United States into the Atlantic provinces (excluding Prince Edward Island and Newfoundland). Eastern cougars likely once ranged farther west through south-central Quebec and Ontario as well, and possibly north of Lake Superior and even into southeastern Manitoba. There are no reliable estimates for historic numbers, but eastern cougars were probably always relatively scarce. There would have been few deer in the then-virgin forests, and this was really a rare cougar population at the northern extremity of its range. Based on the last few authenticated specimens killed, cougars appear to have virtually disappeared from eastern Canada by the late 1800s.

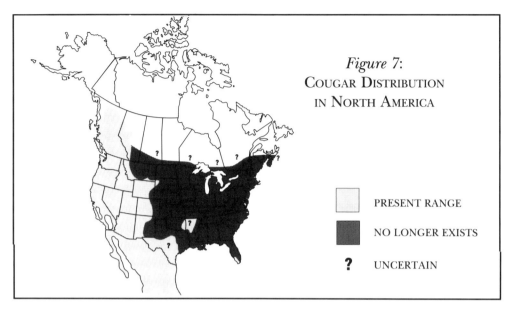

Figure 7:
COUGAR DISTRIBUTION
IN NORTH AMERICA

PRESENT RANGE

NO LONGER EXISTS

? UNCERTAIN

Source: M. Novak, J.A. Baker, M.E. Obbard and B. Malloch, © 1987, Queen's Printer for Ontario (Toronto: Ontario Ministry of Natural Resources and Ontario Trappers Association), p. 658.

From 1900 to the 1940s, there were reported sightings of eastern cougars, many unconfirmed, at an average rate of only one per year. But since then, there have been considerably more. These sightings have been tabulated, recorded, and plotted by various researchers, and a heated scientific and popular debate has arisen over whether people are seeing wild cougars or animals that may have escaped from captivity.

In the 1980s, wwf funded an unsuccessful search for the eastern cougar in Manitoba. Then, in 1990, a particularly exciting but controversial videotaped sighting from New Brunswick was reported on a television news program. The excitement of the story is captured by this lead article from *Panther Prints*, the first official newsletter of a group called Friends of the Eastern Panther:

Wildlife History Is Made in New Brunswick!
The Waasis Panther
Eastern Panther appears in Waasis, and is videotaped alive in its natural habitat for the first time ever. How could a wildlife newsletter have a more auspicious beginning than the opportunity to

report an event of historic importance? Yes, this is exactly what happened on May 15, 1990, in the little village of Waasis, which lies about 12 miles southeast of Fredericton, the capital of the province of New Brunswick. A young panther was spotted moving along the forest edge by Roger Noble, standing at a window in his brother's woodworking shop. At first, unsure of the identity of the animal, he was perplexed. Then, realizing that he was indeed seeing a young Eastern Panther, he called to his sister-in-law and brother.

Then followed approximately ten minutes of exciting action, which culminated in the first-ever filming of a wild, free-ranging Eastern Panther at home in its native habitat, in this case, the far edge of a field bordered by a thick forest of mixed evergreens and hardwood. The cameraman, Roger Noble, had never before used the video camera with which he hurried outside to film the animal. By this time he was well aware of the fact that he was dealing with a panther. Filled with emotions of both excitement and fear, it is remarkable that Roger Noble did indeed manage to come up with a videotape that, while rough and chaotic in parts, nevertheless captures the animal in a variety of modes including walking, standing, sitting, and leaping. Had Roger been thoroughly familiar with the operation of the camera, we would have had an even more remarkable film, including initial sequences taken at a distance of less than one hundred feet. As it turned out, these sequences failed to appear because one camera control had unknowingly been thrown into the wrong position.

We owe a great debt to all three of the Nobles, Donna, David and Roger. Without their great presence of mind and courage, we would still be without hard proof of the Eastern Panther's existence, and this whole incident would have been dismissed by many people, including most wildlife professionals, as just another sighting to be filed away with hundreds of others accumulated over the last few decades. Instead, we can celebrate a historic occasion — the first videotape ever taken of the Eastern Panther (*Felis concolor couguar*, Kerr) alive in its natural habitat.

This videotape has been reviewed by many cougar experts. Some, such as Jay Tischendorf, say it's definitely a cougar. Others aren't so sure. In any case, this sighting has certainly sparked further interest in the wild cat. No doubt many more people will now be looking for it. Also, a panther recovery team will likely be formed by the provincial government, in cooperation with other interested parties, to plan for the protection and recovery of the eastern cougar.

There appear to be adequate deer densities to support a cougar population in about 25 percent of New Brunswick, and areas of high deer density do overlap with frequency of cougar sightings. Therefore, people appear to be seeing cougars where we would most expect them to occur. The 1978 COSEWIC status report notes that, theoretically, a core area in New Brunswick with high deer densities could support 140 to 250 cougars. However, the report also notes that the effects of human settlement and agriculture may be severe enough to rule out the existence of cougars.

This debate is haunted by the fact that we know virtually nothing about the ecology of eastern cougars. We must base our guesses on what we know about the western subspecies. For example, we don't really know why a relic population in New Brunswick, if it exists, hasn't expanded in the last few decades. Is it because the population was low to begin with? Is it because factors other than food availability, such as a lack of suitable cougar habitat, have restricted the population's growth? With a tinge of frustration over this particular question, the COSEWIC report concluded, "In the absence of reliable and detailed information, not much else can be said about the subject." Needless to say, the eastern cougar continues to be formally classified as "endangered," and the debate rages on over whether it still exists or not.

In western Canada, British Columbia and Alberta harbor virtually the entire viable population of cougars in Canada. Here, the species appears to occupy more or less all of its historic range (see Figure 7). Cougar habitat encompasses over 72,000 square kilometers (27,800 sq. mi.) of foothills and mountains in southwestern and central Alberta, with the highest cougar densities reported for the southwest foothills. Best estimates for the current Alberta cougar population are slightly more than 600 animals, with less than 10 percent found inside parks and reserves. Cougar habitat in Alberta joins up with that of British Columbia. In a 1988 paper, *The Status and Management of Cougars in British Columbia*, Daryll Hebert estimated the minimum British Columbia cougar population to be between 2,280 and 3,800 individuals, with the highest densities on

Vancouver Island, and in the Kootenay and Cariboo regions. However, Hebert notes that these are subjective population estimates, and that British Columbia has not as yet undertaken more detailed, quantitative population estimates.

Bounties on cougars existed in British Columbia from 1910 through 1957. In the twenty-five years between 1930 and 1955, 13,257 cougars were killed through hunting and other human activities, which is an average of about 530 cougars reported killed per year during this "bounty era." The cougar was classified as a big-game species in British Columbia in 1966, hunting seasons became more regulated in 1968 and 1969, the issuance of hunting tags to restrict the number of cougars that could be legally killed was initiated in 1970, and compulsory inspection of all cougars killed started in 1976. During this period of compulsory inspection, between 1976 and 1988, the average number of cougars killed through hunting was reduced to 190 animals per year, with a low of 150 in 1981 and a high of 248 in 1986. Females with kittens were legally protected in British Columbia in 1980, and still are.

In Alberta, cougar bounties existed between 1937 and 1964. During this period, the average number of these wild cats killed appears to have been about forty to fifty each year, with a high of over seventy and a low of under twenty. In 1972, a compulsory registration system similar to British Columbia's was introduced in Alberta. Up to 1988, the average number of cougars reported killed by hunters was about thirty per year, with a high of thirty-five and a low of twenty-one. In examining the distribution of this kill, it is clear that most of it (80 percent) occurred in southern Alberta, and that some wildlife-management units (smaller areas used for setting hunting regulations) have been overhunted in some years. As recently as 1990, a cougar quota system was introduced in the province to prevent this overkilling from occurring.

The big picture regarding what has happened to cougars in western Canada is similar to that of other large carnivores. They have moved through a period of intense persecution, bounties, and poorly regulated hunting. Consequently, adjustments have been made, mostly to the government-set regulations concerning the hunting kill. These regulations appear to have at least stabilized cougar population levels in both British Columbia and Alberta, and some may even be increasing. Although the cougar still occupies more or less the same range in the West as it did formerly, it is not known how current populations or densities compare with historic numbers.

It is interesting to note that most of the conservation measures regarding cougars appear to be related to hunting. Of course, human hunting of cougars represents a primary biological pressure on cougar populations, as well as a "user group" for which government, through regulations, has tried to maintain these animals. This should not be interpreted as a "bad" thing. In fact, cougar hunting in western Canada does not appear to be threatening the species; rather, it has been the main interest driving conservation policy. As Ian Ross puts it, "It seems that the only people who get 'tangible' use of cougars are cougar hunters."

Perhaps, since cougars are not frequently encountered by non-hunters, there is not yet a strong "non-consumptive" voice motivating cougar conservation in western Canada. Indeed, most people appear to be relatively unaware that cougars even exist there. Ironically, in eastern Canada, where the cougar may no longer exist, there is a strong protectionist voice for this wild cat of our wilderness, as Canadians who "just want to know that it is there" try to find and save a ghost-like relic cougar population. As Canadian songwriter Joni Mitchell wrote, "You don't know what you've got 'till it's gone" — a disturbing thought we hope will not tell the story of the cougar in the whole of this country.

CURRENT THREATS TO COUGARS

Most of the conservation effort with regard to cougars has been directed at protecting the habitat of its prey, principally deer. As Daryll Hebert from British Columbia reports: "In general, the protection, management, and enhancement of prey habitat and populations are the main sustaining factors for cougar populations in British Columbia, and probably throughout North America."

Cougars live not only in mountain habitat, but in foothill country as well. The aspen parkland and rolling foothills less than 50 kilometers (31 mi.) outside Calgary are excellent cougar country. In fact, cougars are found just about anywhere in southwestern Alberta where there is sufficient cover for cougars to stalk white-tailed deer, mule deer, elk, and moose calves. "If those things are there, cougars seem to be there," observes Martin Jalkotzy. "Because cougars are so secretive, they are living right under our noses and we don't even know it."

Unfortunately, this cougar habitat in Alberta has also become prime real estate, particularly around Calgary and other population centres, as Jalkotzy also notes, "not so much for agriculture, because that's happening everywhere along the eastern fringe of the Rockies, but for subdivisions, golf courses — that sort of thing. We're eating up the boreal foothills that are cougar habitat, and deer, elk, and moose habitat as well. Cougars can't survive in those situations, so they're pushed out . . . it's an incremental loss, not a short-term, but a long-term problem."

In British Columbia, similar pressures are apparent in areas such as the Okanagan Valley. And every year, there are reports of cougars turning up in small towns and even cities such as Victoria, especially where dogs are allowed to run loose and consequently chase these big wild cats. Usually, cougars found in these locations are shot as a threat to people, even though these are clearly situations of people invading wildlife habitat and becoming a threat to cougars. Martin Jalkotzy hammers the habitat point over and over again: "We have to make carnivores and their prey a priority in planning the management of our remaining wildlands. That is really the cornerstone to it. If we don't take into account, at the planning level, what we are doing to our wildlife in general — and our carnivores are a good indication of what's going on — then we're in big trouble. . . . We're getting a lot of good *words* about that kind of thing now, but on the ground it's not happening."

In eastern Canada, if the cougar has, in fact, been able to hang on, its future is entirely dependent on protecting habitat that provides the prey base for this endangered species. Remarking on the tenacity of the Florida cougar, Jay Tischendorf observes, "If they've been able to hold on for three or four centuries down in those fairly marginal swamp habitats, I think there is great evidence that, up in the northeast, in New Brunswick especially, the cats could continue to hold on as well." But it can only "hold on" if we protect the larger natural system or habitat upon which the eastern cougar depends.

The cougar's fear of dogs, and its instinct to climb a tree when chased by hounds, accounts entirely for how this predator is hunted by humans. Its fear of hounds and barking dogs is puzzling because a full-grown adult female cougar weighing 45 kilograms (100 lbs.)

will climb a tree to escape a baying dog as small as a dachshund. Perhaps, in prehistoric times, there was a canine predator that was very dangerous to cats. Once they get up in a tree, however, cougars know it's safe and they virtually go to sleep, slumped down in the branches. Martin Jalkotzy "finds this kind of strange; that a predator as capable and efficient as the cougar will run from these little canids. Oddly enough, these same cats prey on coyotes, usually when a coyote is at or near a cougar kill. And cougars will defend kills from wolves, even killing a wolf that is foolish enough to come too close."

Hunting cougars entails, first, finding cougar tracks, then turning the hounds loose to chase the wild cat up a tree, from where it is shot. Male cougars are the most sought after because, on average, they are about one and half times bigger than females and thus make a better hunting trophy. Nevertheless, historically, in both British Columbia and Alberta, females have made up about 45 percent of the total number of cougars reported killed.

Cougar hunters, although not directly endangering the species as a whole, must face some ethical questions. For example, the non-hunting public often finds it difficult to understand what sport is involved in shooting a cat out of a tree. Unlike moose or deer, the meat of the animal is not eaten; it is simply "dressed out" as a "rug" or a preserved trophy. In addition, it is believed that a few hunters don't even buy a cougar license until their hunting guide calls to advise them that a cougar has been chased and successfully treed. This "will-call" hunt results in the cougar remaining treed by the hounds for a long time while the hunter purchases a cougar hunting license, before finally arriving at the site to shoot it from the tree. High-tech refinements, such as putting radio-telemetry collars on hunting hounds to more efficiently locate the dogs and the treed cat, are also drawing public criticism. Although, from a conservation standpoint, the cougar is just as dead regardless of the strategy or technique used to kill it, it is difficult to argue that such practices constitute ethical hunting in any natural sense of the word.

Although hunting does not appear to be a serious threat to the overall population of cougars in western Canada, there is evidence that overhunting can occur in areas when snow conditions are optimal for tracking. Martin Jalkotzy cited seventeen cases of the annual hunting kill exceeding 20 percent of the cougar population in specific areas in Alberta between 1978 and 1989. That figure was too high. In order to improve this situation, the government has set a limit or quota on the number of cougars that may be killed in every management unit. It is important to close the cougar hunting season when that number is reached, rather than allowing the hunt to continue. Alberta implemented cougar hunting-regulation changes in 1990 that accomplished those ends.

Another threat to cougars related to hunting is the training of cougar hounds in the "off season." This practice has been known to result in wild cats being chased, treed, and generally harassed, even if they aren't shot. Obviously, young kittens could be particularly vulnerable to being chased, and even caught, by cougar hounds under such circumstances.

In eastern Canada, hunting of any cougars would obviously further endanger the population, and that is unacceptable. The eastern cougar presents a clear example of what can happen to our large carnivores if we do not plan for their conservation.

Building roads into cougar habitat has been fatal for cougars, as it has been for many other wildlife species. First, roads access wilderness backcountry and increase human hunting of deer. The result can be and has been a reduction in the number of deer available for cougars, and even the inevitable demands for cougar control in order to make more deer available for human hunting. All because a road was built. Second, because cougars are quite wide-ranging predators, the more roads there are in an area, the more often the cougars cross them, leaving their tracks behind. These now-accessible track crossings are used by cougar hunters as sites to release hounds, thereby making the wild cats more vulnerable to increased hunting pressure. Third, roads generally reduce the wilderness character of an area, and cougars, secretive as they are, select the solitude of unaccessed backcountry. Even more important is "road mortality" — cougars run over by vehicles. Four of seven cougar deaths in a study in the Kootenays in British Columbia were the result of vehicle collisions.

Although province-wide bounties on cougars have been discontinued in Canada, the fact that cougars prey on deer and other ungulate species that are also sought by people leaves them vulnerable to local demands and efforts, both legal and illegal, to control cougar numbers. Cougars are sometimes caught in traps set for other animals, and since it is difficult to release a cougar, it is often killed in these circumstances. Cougars are also easily poisoned, because they will return to a poison-laced carcass, even if human scent is present. The argument for cougar-control programs is sometimes further advanced by the claim that certain prey species, such as bighorn sheep, are already threatened. In such cases, as with demands for wolf-control proposals, all threats to the prey species should be addressed, including habitat loss, hunting, and other human-induced impacts. Finally, since cougars are sometimes a threat to livestock in or near wildlands, and occasionally to humans themselves, calls for widespread cougar control ensue, rather than efforts to deal with the individual problem predator or situation.

Jay Tischendorf raises two other concerns, particularly with respect to endangered cougar populations in eastern North America. The first concern is what he calls "intrusive science." Jay would like to see "field researchers have a deeper appreciation not only for the animals they're studying at the population level, but for individuals as well. Most researchers seem to have a 'we need to know' attitude, and they accept that what they're doing may be detrimental to the individual, but they frequently fall back on this 'carte blanche' that they're helping the population or species. Sometimes that's true; sometimes it isn't. We need to be careful when designing research projects that we're not going overboard, getting carried away with too much intrusiveness, too little concern for the animal. I think that's very important; science does tend to get carried away."

A second concern is people having cougars, or any wild animal, as pets, either in zoos or through private ownership. This is a problem for wolves, too (see Appendix G). It likely has little direct impact on wild populations as large numbers of animals are not taken from the wild for this purpose. However, when captive animals escape, Tischendorf points out that it can confuse the issue of whether or not a wild population actually exists, as in the case of the eastern cougar. Also, because escaped captive animals typically get close to humans, their behavior is interpreted as threatening, rabid, or aggressive, and then the entire species "gets a bad name." Such problems could be avoided if large carnivores weren't kept as pets in the first place. Our conservation goal, after all, should be to maintain fascinating animals in the wild.

BLUEPRINT FOR SURVIVAL

Conservation measures outlined in Chapter 8, which cover all large carnivores in Canada, are important for cougars as well. In addition, however, a number of specific steps apply to cougars in particular.

1. *International Guidance*

The IUCN Cat Specialist Group has produced a statement called *Saving the Wild Cats* (see Appendix E). As might be expected, this document focuses primarily on endangered spotted cats, but it should not be considered irrelevant to Canada or North America. After all, our own large cat is endangered in the eastern part of its former range.

The IUCN Cat Specialist Group outlines general principles regarding why cats should be conserved, problems faced by wild cats, the decline of the big cats, problems of cat conservation, and how wild cats can be conserved. All of these topics are relevant to the cougar in Canada; therefore, this international conservation document should be considered as a guide for a Canadian blueprint for cougar survival.

2. *Canadian Strategy*

Canada has two separate tasks with respect to a national conservation strategy for cougars: one, to maintain healthy cougar populations in the West; and, two, to protect and rebuild endangered cougar populations in the East.

With respect to the non-endangered western cougar population, the national conservation task falls to British Columbia and Alberta, where virtually all of our remaining healthy populations of cougars are found. These two provinces must, therefore, set a goal of maintaining viable populations of these wild cats, based on estimates of their current numbers. Conservation strategies should then be designed at the regional level to maintain the overall numbers of animals, and to protect the Vancouver Island, coastal, and mainland subspecies.

Because their cougar ranges overlap, British Columbia and Alberta should confer regularly to assess the overall status of cougars in Canada, and to make sure that their respective provincial management practices relate to national conservation goals with respect to this species. A joint report in this regard should be submitted annually by British Columbia and Alberta to the annual Federal/Provincial/Territorial Wildlife Conference — an annual gathering of all government wildlife agencies and key non-government conservation organizations in Canada.

The Friends of the Eastern Panther have outlined four objectives, with respect to the eastern subspecies, that would involve international cooperation between Canada and the United States:

• The assembly of sufficient, clear evidence to convince the provincial and state wildlife officials in the northeast that the eastern panther does in fact exist as a viable and stable population.
• The identification, preservation, and enhancement of the animal's breeding ranges.
• The preservation and enhancement of existing eastern cougar habitat, culminating with the establishment of the Bruce S. Wright International Panther Range in New Brunswick and Maine.
• An expanded system of state, provincial, and national parks, with special emphasis on a major westward extension of Fundy National Park in southern New Brunswick.

These objectives reflect what Jay Tischendorf calls the "Five Rs" with respect to the eastern cougar: "We need conclusive proof of its existence — the *recognition* of the eastern panther; we need to find out where its *range* is or where the populations and individuals are; we need to find out how successful the animal is in its *reproduction* and how viable we can expect a future population to be with proper management; we need to figure out the *reasons* why the animal has chosen certain areas for its continued existence; and we need to understand what *requirements* the animal has, so we can base future conservation efforts on these."

The eastern cougar is, or will be, fully protected by the Endangered Species acts of Manitoba, Ontario, New Brunswick, and Quebec. In addition, both the Florida cougar and eastern cougar are listed on Appendix I of the Convention on International Trade in Endangered Species of Wild Flora and Fauna (CITES), which means that international trade in this wild cat or its parts is not permitted by all signatories to the convention (see Appendix D).

3. *Habitat Protection*

Since degradation of their habitat is the biggest threat to both western and eastern cougars, their ecological needs must be factored into any proposed developments in formerly wild backcountry, and even in rural areas. This is especially true for residential developments such as subdivisions and town expansions, recreational developments such as golf courses and tourist lodges, and road building intended to provide access for residential, recreational, and industrial activities. Attention must be placed not only on direct impacts on cougars themselves, but also on their primary prey species, namely, deer. As Martin Jalkotzy has already pointed out, it is one thing to utter the fine words that indicate that such things will be done; it is quite another thing to implement them at the planning level to make sure they make a difference on the ground. When will we finally understand that the status of top predators, such as cougars, indicates how well we are doing in protecting the entire wildlife system, and begin to take these fine words seriously by insisting they are acted on?

4. *Hunting Regulations*

Hunting of cougars in both British Columbia and Alberta should be controlled by quotas set at the management-unit level. Hunting should terminate when the predetermined quota is reached, even if that occurs before the hunting season ends. This practice would also help each province to maintain cougar subspecies, by preventing overhunting at the local level. By making sure regulations prohibit overkilling at this local level, each province will be insuring that its conservation goals for this species are met. Furthermore, since the two provinces have a national opportunity and responsibility for insuring the future of the cougar in Canada, this local management tool would further the realization of that larger Canadian conservation objective as well.

These regulations should stipulate specifically that human-caused cougar deaths should not exceed more than 15 percent of the estimated cougar population each year. Of this 15 percent, recreational hunting should not exceed 10 percent, with no more than 50 percent of the animals killed through hunting being females. Once the maximum number of females permitted is killed from within a particular wildlife-management unit, the cougar hunting season should be closed, even if the total number of cougars permitted (males and females) has not yet been taken. This measure would prevent females, mistaken for males, from being killed accidentally after the maximum number of females has been reached.

Currently, the estimated provincial cougar population in Alberta is about 635, and regional hunting quotas allow for the killing of 58 per year. That means about 9 percent of the cougar population in the province can be legally killed in this manner, excluding "problem" cougars killed, road kills, and accidental mortality. This hunting quota is within the suggested 10 percent hunting guideline. In the first hunting season, when a new system as outlined above was implemented in Alberta in 1990, it resulted in a total of 48 cougars killed, about 7.5 percent of the estimated provincial population. The ratio was 60 percent males to 40 percent females, within the 50 percent guideline for females. These data indicate that the new regulations are more effectively controlling the hunting kill of cougars in Alberta.

In British Columbia, the provincial cougar population is estimated to be between 2,280 and 3,800 individuals, and the average kill is about 190 cougars per year. The annual provincial hunting kill is thus somewhere between 5 and 8 percent of the estimated cougar population. Here, too, about 45 percent of the cougars killed through hunting are females. This appears to be within safe conservation levels, but closer to the upper limit if the provincial cougar population is on the lower side of the population estimate. (Again, this does not include the number of cougars killed as "problem" animals, road kills, and accidental mortality.) This finding points to the importance of doing further research on cougar population numbers in British Columbia. In both British Columbia and Alberta, however, rather than citing province-wide numbers of cougars killed through hunting, it would be more meaningful to examine levels of kill at the local level. In this way, it could be determined whether or not overkilling of a subpopulation — from all causes — is taking place. In some specific areas, this has been a problem.

The following suggestions should be implemented wherever

cougars are hunted, and where these actions are not already enforced. Some, but not all, of these are currently part of the hunting regulations in British Columbia or Alberta:

• Hunters should be required to submit fresh cougar hides and/or skulls to wildlife officials to assist with population census work.
• Cougar kittens and females traveling together should be protected at all times.
• "Will-call" hunts should be ruled out, by requiring all cougar hunters to purchase their licenses by a certain date before the cougar hunting season opens.
• No jurisdiction should have open periods outside hunting seasons when cougars can be run by hounds and treed. Although done primarily to train dogs for the hunting season, this practice could be abused for wildlife viewing or research purposes. It needs to be strictly controlled under special permits and regulations.

5. *Control Programs*
Currently, there are no cougar bounties in Canada, and there should be no future consideration of them. Whenever possible, cougars preying on livestock should be relocated rather than killed. The grazing of sheep, goats, and cattle in cougar habitat should be discouraged, especially in upland areas of British Columbia. A further non-lethal preventative measure could be the use of livestock guardian dogs (see Chapter 5).

6. *Public Information*
Because cougars are so seldom seen in the wild, it is especially important for non-hunting members of the public to demand information on cougar ecology and field signs. Just being aware that you have passed through cougar country because you were able to recognize their tracks, scats, scratchings, or other signs, would heighten a wilderness experience for anyone. It would also garner much-needed support for cougar conservation.

7. *Research*
There always seems to be a long list of things we don't know

about any species, but some specific research topics that are important for the conservation of cougars include:

- developing better population census techniques, and using them to get a more accurate fix on overall cougar populations, particularly in Alberta, British Columbia, and New Brunswick;
- better understanding the habitat requirements of cougars, especially the density of deer required to maintain viable cougar populations, so that residential development, hunting, and other human impacts are not allowed to become so intense as to cause resident cougars to abandon their territories;
- monitoring the age and sex of cougars killed, especially through hunting, to better predict and therefore control impacts on the overall population; and
- studying the effects of competing predators, especially bears and wolves, on cougars.

The situation for the cougar in Canada is quite different for the eastern and the western subspecies. In eastern Canada, the cougar is on the brink of extinction, and the strongest protectionist measures possible must be taken to ensure its future. In western Canada, the conservation key for cougars lies in protecting its wilderness habitat. This measure includes maintaining adequate numbers of deer for food, as well as preserving wild, natural terrain for stalking and other ecological needs. There should be no compromise on habitat protection for cougars in the West, but extremes on the hunting/no hunting debate should be avoided. On the one hand, total protection is not really necessary, given the fairly healthy population status of cougars in British Columbia and Alberta. On the other hand, cougars are relatively slow-reproducing, and can be overhunted if careful regulations aren't maintained. Furthermore, we are still uncertain about the size of cougar populations, particularly in British Columbia and northern Alberta. Therefore, caution is in order when considering the number of cougars that can be taken each year out of any cougar population without causing long-term declines.

Finally, it will be interesting to see whether a stronger voice emerges on behalf of the cougar from people who may never see one, who are not hunters, who might like to see signs of cougars, or who simply want to know cougars are out there. Whatever the motivation, our own large wild cat deserves conservation attention just as much as its famous spotted, striped, and long-maned relatives elsewhere in the world.

7.
The WOLVERINE

Probably no other northern mammal plays such an important role in

camp-fire tales and folklore as does the wolverine. It is accredited with

prodigious strength and ferocity — enough to drive off a grizzly bear — and

superhuman intelligence. . . . The wolverine is pugnacious,

curious, bold and strong.

A.W.F. BANFIELD, FROM *THE MAMMALS OF CANADA*

THE WOLVERINE IS THE STUFF OF LEGENDS AND MYSTERY. SCIENTISTS KNOW little about its basic biology or behavior. Its scientific name, *Gulo gulo*, was derived from the Latin word *gulosus*, which means "gluttonous." It is seldom seen because it is found in very low numbers, travels huge distances over remote areas, and, although it is believed to be most active at night, in fact no clear activity patterns have been established for this animal. When it is encountered, chances are the wolverine may have robbed a trapline or ransacked a wilderness cabin in search of food, demonstrating its unusual intelligence or sheer strength. All of these characteristics help explain why so many camp-fire tales — strange mixtures of fact, myth, exaggeration, loathing, and awe — float around about the wolverine. When you don't know much about a wild animal, but what you do know inspires these kinds of responses, then it's easy to fabricate the rest and conjure up a "devil beast" — one of its nicknames.

WOLVERINE FACTS AND FIGURES

Wolverines are actually large members of the "mustelid," or weasel family, not of the "canid" or dog family, as some people believe. In fact, the wolverine is the largest weasel living on land in Canada. Only the sea otter, another member of the weasel family, is bigger. In many respects, the wolverine looks like a small bear — an old legend says that, if a female bear has four cubs, the fourth will be a wolverine. It is a compact, muscular animal with a broad head, relatively short strong legs, and large, well-clawed feet, which equip it to travel over snow, dig for carrion, and even climb trees. It is the smallest of our large carnivores, with a body length ranging from 65 to 105 centimeters (2 to 3.5 ft.), and a medium-sized bushy tail 17 to 26 centimeters long (7 to 10 in.). Wolverines weigh between 14 and 21 kilograms (31 and 47 lbs.), with males usually weighing about 30 percent more than females. In his book, *Mammals of Canada*, A.W.F. Banfield vividly describes the wolverine as "about the size of a fat spaniel dog or a bear cub."

In color, the wolverine's furry coat is usually dark brown with two pale buff-colored flank stripes starting at the shoulder, extending down both sides of the body and joining at the base of the tail. These stripes are one reason why the wolverine is often nicknamed "skunk bear." Most wolverines also show a light facial mask, contrasting with beady black eyes and a dark muzzle. They have white or blondish fur along the throat and chest, and their short rounded ears are often tipped with a grizzled gray fur. All these colors vary according to individuals — some being reddish or medium brown overall, to almost black, with the light markings ranging from nearly white to a blondish color or tan.

The fur of the wolverine is unique and prized for its "frost-free" qualities. Ice crystals may form on the underfur, but not on the long guard hairs, whereas wolf or coyote fur tends to become matted with ice. When frost from a person's breath forms on wolverine fur used for trim around the hood of a parka, the frost is easily brushed or shaken away. For these reasons, this animal's fur is particularly valued by aboriginal people, who have long known about these special characteristics.

The wolverine has extremely strong jaws and teeth — something like the hyena's — which allow it to crush the bones and frozen meat that it often digs up from carcasses under the ice and snow. Its sight isn't particularly sharp, but its sense of smell is well developed, enabling wolverines to locate carrion as much as 2 metres (6.5 ft.) beneath the snow. Their acute sense of smell also means they can detect the scent of humans while we are still far away. Wolverines have usually left an area long before a human visitor reaches it, although some enterprising ones have learned to associate food and people and are attracted to the human scent.

Wolverines have anal musk glands from which they release a pervasive-smelling, yellowish fluid to "mark" their food, making it undesirable to other animals and humans. Hence, a second reason for the name "skunk bear." This scent can be found on carcasses the wolverine hasn't finished eating, on food caches to which it plans to return, and on the inside of cabins, very clearly indicating that the wolverine has broken in! Releasing this scent is considered by some researchers to be a marking strategy, perhaps indicating owner-ship or territoriality. The animal may also secrete this musk if alarmed or stressed.

Like the other top predators, wolverines can be divided into dif-ferent subgroups according to the habitats where they are found. Generally speaking, wolverines frequent four habitat types. The Pacific Forest Type includes wolverine range along the coast of Washington, up through British Columbia to southern Alaska, for about 150 kilometers (94 mi.) inland. The Rocky Mountain Forest Type includes wolverine range through Colorado, Montana, southwestern Alberta, and the interior of British Columbia. The Boreal Forest Habitat Type encompasses the largest part of the wolverine's range, including part of Alaska, the southern Northwest Territories, Yukon, Alberta, some parts of British Columbia, and the northern part of eastern Canada. And, finally, the Tundra Type makes up the most northerly part of the animal's range, especially in the Northwest Territories, and the far-northern parts of eastern Canada.

Within these four habitat types, two wolverine subspecies and one subpopulation are of conservation concern. Although scientists disagree on whether or not distinct subspecies even exist, some believe there may be a distinct Vancouver Island wolverine subspecies (*Gulo gulo vancouverensis*), a smaller wolverine with a chestnut hue to the flank stripes. If this is a separate subspecies, it is geographically isolated from the mainland and extremely rare. The other subspecies is the western wolverine (*Gulo gulo lutens*), occurring only in southwestern British Columbia, Washington State (excluding the southeastern corner), the western half of Oregon, and parts of California. This subspecies exists in high mountainous elevations. Its range has become fragmented, or broken up into "islands." Therefore, it, too, demands special vigilance if its future is to be assured. The subpopulation of wolverines that is of greatest conservation concern is comprised of all wolverines still existing east of Hudson Bay. In 1989, the "eastern wolverine" was officially classified as "endangered" by the Committee on the Status of Endangered Wildlife in Canada (COSEWIC). Thus, the eastern wolverine could become extinct if the factors affecting its decline are not reversed.

As with the other top predators discussed in this book, it is important to design conservation strategies that maintain not just "the wolverine" in Canada, but wolverines in all these various habitat types, as well as possible subspecies and subpopulations.

Conservation strategies must be based on a species's biology, and the wolverine's biology, more than that of any other top predator, remains somewhat of a secret for now. But this much is known: male and female wolverines become sexually mature in their second year. However, the female has an interesting biological characteristic called "delayed implantation." Even after the female wolverine has become fertilized by a male, she may or may not fully develop the embryos to give birth to kits. The fertilized egg goes into a "blastocyst" stage (a small ball of cells), and can stop developing at that point. The blastocyst floats around in the female wolverine's uterus until sometime between November and January, when it implants into the uterine wall and further develops into kits. Wolverine kits are then usually born between February and March.

However, if conditions are not good during the blastocyst stage or at the time of implantation — for example, if food is scarce — the female wolverine's body may reabsorb the embryo. In effect, she terminates the pregnancy, and will have no young that spring. Delayed implantation, therefore, allows the female to be bred and become "officially pregnant" in the summertime when food supplies are good, but she maintains the option of not having young, depending on winter conditions. As an example, an Alaskan study showed that female wolverines did not have litters every year when food supplies were low, even though they may have bred. Therefore, although female wolverines come into season once a year after they've reached the age of two, and even if they breed every year, they may not necessarily give birth at yearly intervals. In another study, in the Pacific United States, wolverines produced litters only every second or third year. In Alaska, only about one-third of the adult females were observed with young every year.

Litters range from one to five kits, but the average is two or three. Certainly not all of these survive to become reproducing adults, though precise mortality rates are unknown. Newborn wolverine kits are covered with fine white fur. They are only 12 centimeters (5 in.) in length, with stubby little tails, and they weigh about 90 grams (3 oz.) each. Banfield mentions that the mother licks the kits' tummies to help them digest their food, and like the cougar, may attack intruders when her young are very small. Kits grow rapidly, however. They are weaned at seven to eight weeks, are out of the den at twelve to fourteen weeks, and grow to their adult weights by late fall. Young wolverines stay with their mother for their first winter to learn to hunt, then they disperse to lead lives of their own the following spring.

Since researchers believe that wolverines normally live eight to ten years in the wild, the above information regarding reproduction would indicate that the average female wolverine reaching sexual maturity at two years of age, then successfully raising two kits on average every two years, would produce only six to ten young in her lifetime. Scientists are not sure how many of these might survive to adulthood. Depending upon food conditions, reproductive output of wolverines could obviously be more or less than this. Based on the wolverine's known biology, however, these calculations serve to show once again the familiar pattern exhibited by most large carnivores: they have relatively low reproductive rates compared to other animals, which should always be of conservation concern.

Vivian Banci conducted the first wolverine studies in Canada, in the Yukon between 1983 and 1985, with WWF support. Regarding wolverine reproduction, Vivian observes, "If you look at indices of reproduction, such as the number of young, they tend to be rather high for an animal that size. But then, if you look at the productivity of the population, how many young actually survive? It's really low. So, there is something happening between the time they get pregnant to when the young, say, leave their dens. In my study area, which happens to be in some of the best-quality habitat in the Yukon because of its high diversity of prey, I have observed 50 percent of my females successfully producing young. So, there is something limiting the success of reproduction, and that has major implications for the entire population." This crucial conservation mystery remains unsolved.

The best thumbnail description of the wolverine while feeding would be "scavenging predator." Although wolverines will occasionally eat fruits, berries, and insects, by far the majority of their diet is meat. And the majority of that meat comes from the carcasses of animals killed by something else — other predators, disease, starvation, or harsh weather. The carcasses are usually those of ungulates such as elk, deer, and moose in the southern part of wolverine range, and caribou in the North. This finding has led many researchers to suggest that two things are necessary for a healthy wolverine population: first, an abundance of diverse prey, including small mammals, birds and eggs, fruit, as well as ungulates and other hoofed mammals; and, second, a healthy population of predators (especially wolves) to kill enough of the big prey to leave carcasses behind for the wolverine. For this reason, wolverines have been observed following wolf packs in order to scavenge their kills; and wolverine populations have declined in areas where both their prey and other predators have declined.

Once a carcass is found by a wolverine, it is nearly always "marked" by secretions from its musk glands. If there is more food than the wolverine needs immediately, it will often be "cached" — buried in the ground or snow — for later. Researchers in Alaska, who were puzzling over why they were finding large numbers of summer ground squirrels in the stomachs of wolverines in the middle of winter, concluded that the squirrels were caught by the wolverine in summer, cached, and then eaten in winter. Everything, from foxes, ptarmigans, caribou pieces, and ducks, to carrion of all types, has been found cached by wolverines. Since the wolverine is an excellent climber, some of its food caches may even be hung up in tree branches. In the North, people cache food, too. However, the wolverine's acute sense of smell often leads it to these caches, and, with its great strength, it can roll back even heavy rocks covering them. Then, of course, the wolverine's "marking" habit renders such caches inedible for people.

Vivian Banci relates an interesting story about the wolverine's food habits and great strength: "I had horse quarters out as bait for a wolverine I was trying to catch, and I had chained them to a tree. The wolverine just pulled them off! They're so strong. Their jaw muscles and their neck muscles are massive, and their teeth are really robust as well. If you think about what wolverine do for a living, they are mainly scavenging and feeding on frozen flesh and bones. They'll actually eat bone, and they'll grind it with their teeth. I have found stomachs just full of ground bone, like a powder."

Wolverines will also kill live animals, given the chance. Although they are extremely strong, wolverines are not particularly fast. A running wolverine could likely be overtaken by a wolf or lynx, but it would be an exceptional human who could do so. It is thus unlikely that wolverines kill many large animals such as moose or deer. Nevertheless, there are various reports in the scientific literature and in folklore of wolverines doing this. One report that occurs over and over again is that of a wolverine killing a moose. We were fortunate to get the "story behind the story" from Ian McTaggart-Cowan: "Frank Wells was the warden on the Sunwapta River [Alberta] for many years. He told me that he was following the trail of a moose in the wintertime, and all of a sudden he realized that there were tufts of hair on the snow on either side of the trail of the moose, and the moose was running. Then there were blood spatters on the snow, and he came upon the carcass of the moose. It had been killed, and all around it were the tracks of a wolverine. His interpretation of what he saw was that a wolverine had been in a tree, leapt onto the back of the moose, and had simply been hanging on, chewing away at the neck as the thing went along. Now that's the only record I have of a wolverine attacking so large an animal. He didn't actually see it; he deduced it. It could have been a very sickly moose, on its way out. Certainly there was no doubt about what was eating the moose; it was a wolverine."

It is likely that the wolverine supplements its carrion diet with more manageable prey — marmots, chipmunks, rabbits, mice, voles, grouse, birds' eggs, fish, and almost anything else it can find. In her Yukon study, Vivian Banci examined many wolverine carcasses provided by trappers, and found remains and hair of beaver, mink, marten, ermine, lynx, coyote, and even wolf in the stomachs. (The wolf hair likely came from the site of another animal killed by wolves.) Wolverines are also known to eat porcupines, though it's a messy business, often leaving the wolverine with quills in its muzzle and stomach, which can lead to death.

Wolverines are renowned for stealing bait and trapped animals from traplines, and even hiding traps. They have also gnawed their way through virtually everything in order to get into trappers' cabins in search of food. Once in, they chew up furs, traps, and supplies and scent-mark the inside of the building to the point where it becomes uninhabitable. Some old-timers claim wolverines are smart enough to know when bait has been laced with poison, such as strychnine, when the trapper tries to even the score.

The conservation significance of the wolverine's food habits is clear. First, they need carrion, which means wolverines require habitat that produces a diversity of prey, including ungulates, plus other predators to leave carcasses behind. Second, the wolverine travels long distances and has immense home-range requirements because it has to search widely for carcasses. Third, harsh winters can actually be a blessing in disguise for wolverines because more prey die off as a result of starvation, thereby providing more carcasses for wolverines to scavenge. Fourth, if there aren't enough carcasses, wolverines experience lean times, and their numbers can decline as a result of starvation or slowed reproduction in response to food shortage. Fifth, its habit of robbing traplines and raiding cabins has brought the wolverine into conflict with people, which, in some countries, has led to bounties or other broad-based efforts to eliminate it.

WHAT HAS HAPPENED TO WOLVERINES?

The wolverine is known as a "circumboreal" species, which means it is found in boreal-forest habitat around the globe, including Norway, Sweden, Finland, the USSR, Mongolia, China, Canada, and the United States.

In Eurasia, the wolverine appears to have lost only about 10 percent of its historic range from southern Sweden and Finland, from northern Germany and Czechoslovakia, from eastern Poland, from about 1,000 kilometers (625 mi.) of its former range in the southwestern USSR, and recently from the southeastern border of its range in the Soviet Union. Wolverines are protected in four major nature reserves in the Soviet Union, ranging from 10,000 to 350,000 hectares (25,000 to 864,000 acres). The IUCN Species Survival Commission cites wolverine population estimates in Norway at 118 to 183 animals, with a stable population; in Sweden, somewhere between 75 and 100 wolverines exist in a protected population; in Finland, there is a maximum of 83, in a declining population; in the Soviet Union, there are at least 7,000 to 7,500 wolverines; and, in China, the wolverine is very rare — only 17 have been reported in Chinese hunting statistics since 1949.

Generally, this predator's range in Eurasia has shrunk northward, for a number of reasons. Wolverines have been persecuted under a bounty system as a result of their preying on domestically raised reindeer and their breaking into huts. Snowmobile traffic in wolverine refuges has severely disturbed their habitat. And, finally, wolverines have picked up poison bait that was put out to kill wolves.

In North America, the wolverine has disappeared from almost half of its original range (see Figure 8). The greatest loss has occurred in the lower United States, where wolverines are no longer found in Minnesota, Nebraska, North and South Dakota, Utah, Nevada, Indiana, Iowa, Wisconsin, Ohio, Maine, Vermont, New Hampshire, New York, Pennsylvania, New Mexico, and Michigan. This long list speaks for itself.

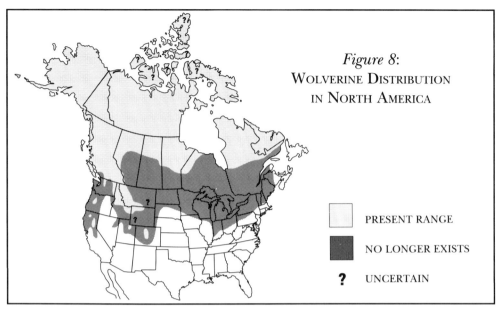

Figure 8:
WOLVERINE DISTRIBUTION
IN NORTH AMERICA

PRESENT RANGE

NO LONGER EXISTS

? UNCERTAIN

Source: M. Novak, J.A. Baker, M.E. Obbard and B. Malloch, © 1987, Queen's Printer for
Ontario (Toronto: Ontario Ministry of Natural Resources and Ontario Trappers
Association), p. 576.

Population estimates for wolverines are very difficult to obtain,
and are usually based on records of the number of wolverines killed
by trapping. It is known that the largest wolverine population in the
United States lives in Alaska, where, in the 1970s, between 500 and
1,000 animals were trapped or hunted every year. The largest, and
only flourishing wolverine population in the lower forty-eight American states is found in Montana, where about 200 per year were
killed during the 1960s. This population represents an increase in
wolverine numbers in Montana, where, by 1920, the wolverine was
thought to no longer exist. It likely recovered there as a result of
wolverines moving down from Canada via Waterton Lakes and
Glacier national parks which have served as "corridors" for recolonizing animals. Wolverine populations in Wyoming, Washington,
Oregon, and California are very small, and it is doubtful they exist
at all in Idaho and Colorado. U.S. national parks that currently have
wolverine populations include Denali and Glacier national parks in
Alaska, and Yellowstone, Grand Teton, Yosemite, and Mount Rainier
national parks in the West. This overall picture has led the IUCN
Species Survival Commission to state that, "in the lower United

States, wolverines are scarce everywhere . . . and their continued survival in the southern Rocky Mountains is uncertain."

In Canada, the wolverine was classified in 1982 by COSEWIC as a "rare" species across the entire country. The word "rare" was subsequently changed to "vulnerable" by COSEWIC. This designation reflects the natural biology of the species, rather than any specific negative impact by people, because the wolverine is naturally found in low numbers over a large area. However, in 1989, all wolverines east of Hudson Bay were declared "endangered," which means the activities of people have brought them to the brink of extinction in that part of the country. In addition, the IUCN Action Plan expresses concern about the two possible western subspecies. The same report states that the wolverine situation in Ontario is "uncertain." Therefore, although its numbers appear to have held up in the northwest, a huge loss of wolverine range has occurred in southern and eastern Canada (see Figure 8).

Despite the fact that it continues to be one of Canada's rarest mammals, the wolverine may nevertheless be more abundant here than anywhere else in the world. The IUCN Action Plan cites wolverine population estimates in excess of 5,000 throughout British Columbia, and over 4,000 in the Yukon. No estimates were available for the Northwest Territories, which likely harbors about the same number of wolverines as the Yukon, and therefore forms the third stronghold for this species in Canada. As for the rest of Canada, the wolverine is no longer found in the aspen parklands of southern Alberta, Saskatchewan, or Manitoba; over 80 percent of this habitat type has been lost from Canada as a result of agriculture and urbanization. Consequently, wolverines are now confined to the boreal-forest regions of those provinces, and to the Rocky Mountains in the case of Alberta. A small population of 70 to 100 animals is thought to be holding on in Ontario, west of James Bay and north of 50 degrees latitude. The IUCN Action Plan mentions small "isolated populations" in northeastern Ontario, northern Labrador, and Baffin Island. High Arctic wolverines are likely very scarce — perhaps just a few animals searching for food by wandering over the sea ice. Wolverines were never found in Nova Scotia or Prince Edward Island, they have disappeared from southern Quebec and New Brunswick, and whether they ever existed in Newfoundland is not certain. National parks in Canada that have wolverines include the Rocky Mountain parks of Jasper, Banff, Kootenay, Yoho, and Waterton Lakes, as well as northern parks such as Kluane, Northern Yukon, Nahanni, and Wood Buffalo.

If we stand back and try to understand the big picture with respect to what has happened to wolverines, we see an overall pattern similar to that of other top predators. Its natural range is inexorably shrinking as the wolverine disappears from its former territory in the South in both North America and Eurasia. In Canada and the United States, this animal has also been pretty well lost from the East, and therefore is taking its last stand in the West, particularly in British Columbia, the Yukon, the Northwest Territories, and Alaska. Here, the wolverine is rare, but likely still more or less as numerous as it has always been. That much is clear.

We are much less clear on why the wolverine has undergone such dramatic range shrinkage. Generally speaking, it appears to be because it is such a vulnerable, thinly spread species to start with, which avoids most aspects of human intrusion and civilization. Therefore, as people invade its wilderness habitat, the wolverine usually moves out. It is no coincidence that reduction of wolverine range in North America appears to have begun around 1840, which coincides with the period of extensive exploration, settlement, and exploitation of wildlife through the fur trade, as well as extermination of species such as the buffalo, wolf, and grizzly bear. As scavengers, wolverines were also heavily impacted by the large-scale, anti-predator programs through ingesting poisons from bait and poisoned carcasses. All of these human activities were impacting on the wolverine, which V. Bailey, in 1926, described in his *Biological Survey of North Dakota* as "an animal of the solitudes, shunning human occupation and vanishing with the spread of human civilization."

CURRENT THREATS TO THE WOLVERINE

Wolverines require large home ranges for their scavenging lifestyle of traveling vast distances in search of food. Virtually everyone who has studied these predators marvels at the distances they travel. For example, Maurice Hornocker, who conducted the key work on this species in Montana, notes that most other animals he has worked with in the northern Rockies are confined to a river drainage, or to a certain portion of land during winter and deep snow. "But not wolverines. They go from one river drainage to the other, right over the tops of these high mountain ranges. It's just phenomenal!"

Vivian Banci, in her Yukon study, also noted that some wolverines were transients, wandering all over. Others would have a large home range and then, inexplicably, move 100 to 200 kilometers (63 to 125 mi.) away for a few days before returning. Vivian wonders, "What are they doing? Maybe all of a sudden something comes into their minds, that they have to go to a certain place — maybe it's an old kill site. For some reason, they have to go, and then they come back. It's like a holiday they take!"

Whatever the reason, wolverine home ranges have been documented to be immense, as large as 963 square kilometers (372 sq. mi.) in Montana, up to 770 square kilometers (297 sq. mi.) in Alaska, and 400 square kilometers (154 sq. mi.) in the Yukon. Vivian lost the radio signal from a transmitting collar on one male wolverine and scoured a 3,000 square kilometer (1,158 sq. mi.) area by plane for a month before he returned! These observations paint the picture of a wild animal that, because it is thinly spread over such big areas to start with, may be particularly vulnerable to human impacts such as direct killing through trapping and alteration of its natural habitat through such activities as forestry, road building, and people settling in wilderness backcountry.

Although wolverines have been the intended and unintended victims of vermin-poisoning and bounty programs in North America, most direct killing by humans takes place through trapping. Because the animal is so wide-ranging, most wolverine trapping is incidental, which means a trapper may take one or two wolverines a season in the course of trapping other animals. Since they are always scavenging for carrion, wolverines are particularly susceptible to traps that use bait to lure wildlife into them.

It may seem obvious that trapping a rare species would cause a population decline. However, this depends on whether the rate of trapping exceeds the species's capability to reproduce. In the case of the wolverine, it is not at all clear that trapping has caused such a decline. For example, the COSEWIC report on this species reported that "the number of wolverine pelts traded annually in Canada has been reasonably constant through most of this and the last century." But we would normally expect a marked decline in the number of wolverines trapped, if the species was declining in the wild. There does appear to have been a reduction in the number of wolverine pelts traded in Canada between 1821 and 1905, when the yearly average was 1,192, versus the period between 1921 and 1985, when the average was 708. However, despite this finding, the 1988 COSEWIC report concludes that "the harvest pattern over the past 150 years does not suggest initial high exploitation leading eventually to near or complete decimation as occurred for the beaver and a number of game species (antelope, bison, wild turkey); rather, it suggests a sustainable harvest from a small, stable population."

Some other aspects of trapping wolverines are worth noting. For example, not all wolverine pelts enter into commercial trade. About 50 percent of the pelts are kept in local communities to be used privately in the Northwest Territories, about 12 percent in Yukon, and less than 5 percent in Alberta. The COSEWIC report estimates that only about one-third of the wolverines trapped in Canada every year enter into legal international trade, and there are no Canadian records of illegal trade in wolverine pelts or parts.

In the eastern part of the wolverine's range, trapping appears to have had a more serious effect on this predator, perhaps because access by trappers into wolverine country is somewhat easier in the flatter, forested lands and tundra of the East than in the more mountainous West. Eastern wolverine populations are naturally smaller than in the West, with or without trapping. And caribou populations in the East have fluctuated dramatically, also likely accounting for declines in wolverine numbers over the past fifty years. The COSEWIC report notes that "many years' harvest could slowly erode the wolverine population on its fringes, especially if habitat quality were also declining. This appears to be what has happened east of Hudson Bay."

It is difficult and risky to use trapping statistics as reliable indicators of a species population being in either good or bad shape. Upward or downward trends in the numbers of animals caught can be affected by a change in the price of pelts, how many trappers there are, the number of traps set, or new trapping technology. Therefore, the trapping statistics may not reflect how many animals are actually "out there." On balance, this discussion of trapping should not lead us to assume that wolverine populations are doing fine, or that they are unaffected by trapping. Rather, there is need to exercise conservation caution wherever wolverines are trapped, and none should be trapped where they are endangered.

Since different populations of wolverines live in different habitat types, from the Pacific forests to the Arctic tundra, maintaining different subspecies or types of wolverines also presents the challenge of protecting a broad range of habitats. In addition, the same individual wolverine, as it ranges over a very large area, scavenging for carrion, will cover many different kinds of habitats, from river valleys to mountain ridges and talus slopes. Therefore, maintaining habitat for even one wolverine presents the challenge of protecting a wide diversity of habitats. Vivian Banci makes the point this way: "They're the quintessential symbol of wilderness . . . there are many factors that are important in order to have wolverines in there, and they're not little factors. They're big ones. As soon as we start thinking big, because they need big wilderness areas, then we can understand how we're going to keep them there."

In forested habitats, timber harvesting and road building have affected movements and behavior of wolverines. Such activities can also affect their food supply, but not always negatively. In Montana, for example, Maurice Hornocker has found that timber operations that removed the tree canopy created a new plant complex in mid-aged stands. This new growth increased the number of rodents available for summer food for wolverines. "But," he emphasizes, "rampant clearcutting of large tracts simply caused wolverines to abandon the area altogether. So the clearcuts have to be small, and they have to be cut with consideration of the aspect and slope. Also, corridors of cover must be maintained between clearcuts."

Hornocker has learned that roads have had an even greater impact than logging. Wolverines will just not tolerate the unregulated human access and disturbance roads bring. "Wolverines are a wilderness animal; they don't like people, and they don't like to be around people," Hornocker says. "Roads not only create summer access for all kinds of recreationists, but winter access with modern snow machines that are capable of going almost anywhere."

The wolverine is a species that, in many ways, has fallen between the cracks. It has been largely unknown in the scientific sense, and its far-ranging nature makes it difficult to "manage." For these reasons, it just hasn't been a priority for most government wildlife-management agencies. As a result, very few conservation measures were even considered for this species until the 1960s. Even today, not many jurisdictions have developed specific wolverine-conservation plans, or even hunting and trapping objectives, because wolverine populations are so difficult and expensive to study.

Vivian Banci has witnessed the truth of this in British Columbia, where, she says, "our ministry has a referral system so that whenever there's a planned land-use activity, such as oil exploration or forestry activities, our wildlife branch has the chance to comment on the impacts for wildlife. . . . Not once has there been a referral where people consider those sorts of impacts on wolverines. And I think that's a great tragedy. Just because you can't see them doesn't mean they're not there. It doesn't mean they don't need habitat, and it doesn't mean they're not important. In fact, in terms of biodiversity and keeping the whole complement of species, especially because they use such a wide range of different habitats in what we typically think of as wilderness, the wolverine should be one of the most critical animals to think about."

In a sense, this neglect is itself a threat to the wolverine. It means we undertake human activities, whether it's trapping, forestry, or road building, in wolverine habitat without really understanding their effects. It means we learn about the effects of such things afterwards, by witnessing further shrinkage in this rare animal's range. However, there are signs that this is changing. Some jurisdictions are expressing new-found interest in the wolverine. For the first time, they are hiring specialists, conducting studies, and putting laws on the books that could make a difference for the wolverine, one hopes before it is lost from still more of its original home.

BLUEPRINT FOR SURVIVAL

The conservation measures outlined in Chapter 8, which cover all large carnivores in Canada, are important to the wolverine as well. In addition, however, a number of specific steps apply to wolverines in particular.

1. *Canadian Population Goals*
   Each jurisdiction in Canada that has not already done so should establish population targets for its wolverines, establish safe harvest regulations, and confer with other jurisdictions regularly

to assess the national status of this animal. Such a national review should be coordinated every three to five years by COSEWIC, especially given the endangered status of the wolverine east of Hudson Bay.

British Columbia, the Yukon, and the Northwest Territories collectively harbor about 75 percent of the Canadian wolverine population. Therefore, they have the greatest opportunity to maintain still-viable numbers of this rare and vulnerable carnivore. Other jurisdictions, particularly Quebec, have a different kind of responsibility, namely, to hang on to an endangered species — the eastern wolverine. Consequently, complete protection must continue there.

2. *Trapping*

Today, the wolverine in Canada has been legally classified as a "furbearer" by all provinces and territories. Therefore, the government agencies responsible for managing wildlife have the responsibility to regulate hunting and trapping, although this legal power may not apply to aboriginal peoples. Table 7 summarizes the legal status and protection for the wolverine in Canada.

It is impossible to calculate what percentage of the total wolverine population is taken every year by trapping in Canada, since population estimates are either non-existent or very shaky. We do know that trapping in Canada resulted in an average of 925 wolverines killed each year between 1980 and 1986. The largest number of wolverines killed were from the Yukon and British Columbia, with an average of 240 wolverines per year each, and the Northwest Territories, with an average of 213. Therefore, these three jurisdictions account for about 75 percent of the total trapping kill. Banci estimates that the number of wolverines trapped in the Yukon represents a maximum of 6 percent of that territory's wolverine population. This percentage may apply roughly to British Columbia and the Northwest Territories as well, but that's uncertain. So little is known about wolverine biology that most wildlife managers are also not sure whether this is a safe level of kill or not.

In fact, trappers have often been a primary and valuable source of information regarding wolverines. They provide their observations to wildlife managers on animal movements, abundance of

Table 7: LEGAL STATUS AND PROTECTION FOR THE WOLVERINE IN CANADA

Province or Territory	Legal Status	Protection
Alberta	Furbearer	Registered trappers only; seasons and quotas
British Columbia	Big game animal and furbearer. Protected on Vancouver Island	Regulated trapping seasons and ability to impose quotas if necessary.
Labrador	Totally protected	—
Manitoba	Furbearer	Regulated trapping seasons and licenses
Northwest Territories	Big-game animal and furbearer	Seasons and zones
Ontario	Furbearer	Harvest monitoring and trapping localities noted. Ability to set quotas
Quebec	Totally protected	—
Saskatchewan	Furbearer	Trapping season
Yukon	Furbearer and game animal	Regulated trapping season, no limits. Limit of one per hunter

Source: C. Dauphiné, *Status Report on the Wolverine (Gulo gulo) in Canada.* Prepared for the Canadian Scientific Authorities for CITES (Ottawa, 1987), p. 22.

carcasses, etc., and they supply wolverine carcasses to researchers trying to assess the health and age composition of a population. This information is especially critical, considering the secret nature of the wolverine and the difficulty in gathering information on such a rare species, spread out over diverse habitats.

With respect to trapping, specific regulatory measures that have been implemented elsewhere should be considered by all Canadian jurisdictions. They include:

• closely monitoring levels of trapping and adjusting quotas accordingly, since wolverines are a rare species to start with;
• insisting that a seal or tag be attached to every wolverine pelt, to insure trapping quotas are being observed and to better understand the size and distribution of the trapping kill;
• closing hunting and trapping seasons in late winter and early spring to protect female wolverines with kits;
• placing a zero wolverine quota on traplines, or working out cooperative arrangements with native trappers, where the wolverine is regarded as endangered; and

- prohibiting bait-trapping of any wildlife species in the areas where the wolverine could expand its range, or where it needs additional protection.

3. *International Trade*
The wolverine is not protected by the Convention on International Trade in Endangered Species of Wild Flora and Fauna (CITES), so there is no monitoring or restriction of international trade in the wolverine or wolverine products. If the proportion of Canadian wolverine pelts in international trade reaches the point where it exceeds 50 percent of the total pelts taken, the wolverine should be listed on Appendix II of CITES, to provide for more extensive monitoring or restriction in trade where needed (see Appendix D).

4. *Habitat Protection*
When considering the impacts of human, industrial, and recreational activities on, and access into, wilderness areas where wolverines are still found, there must be provision for assessing the impacts on this species. Such intrusive proposals may then be modified to mitigate negative impacts on wolverines and their habitats, or disallowed if the impacts on wolverines are unacceptable. Nothing less will insure the future of such a vulnerable animal where humans and wolverines come into contact.

Vivian Banci believes this is very important: "People in the resource industries have not, until very recently, considered species like the wolverine and other furbearers as wildlife. When we say 'wildlife' and 'habitat,' they think of deer, moose, and other such species. But what is good for these species is, in most cases, not the best thing for furbearers. They may be smaller, and they may not be hunted for food, but they're still important. That is changing slowly, but it's not happening quickly enough."

5. *Reintroduction*
In areas where their prey-base still exists, where habitat is sufficiently remote from humans and protected, and where wolverines were once found or are now only occasionally present, reintroduction programs are worth trying. Healthy wild wolverines

caught from similar areas where they are more abundant should probably be released in a ratio of two to four females for every male because males are much more far-ranging than females. If they were available in this ratio, the males could still find and breed with the females in the new territory. Males also tend to be more vulnerable to bait trapping, which is why it would be particularly important to prohibit this practice in areas where wolverines are being reintroduced. The fact that wolverines have shown an ability to reoccupy former range is encouraging, and should be further supported.

6. *Research Needs*

The wolverine continues to be the least known of the large carnivores in North America. Until the last ten to twenty years, very few systematic field studies were conducted on this elusive animal. There is still much to learn. Topics particularly important for conservation purposes are:

- developing cost-effective population censusing techniques to better estimate how many wolverines there are in the remainder of their range;
- better understanding what controls wolverine reproductive rates, especially so we can make sure that human killing through hunting and trapping is not excessive;
- better understanding the impacts of people's activities in wilderness on wolverines, so wildlife managers know how and when these activities should be modified to protect the animal, and when we just shouldn't do certain things at all. This really means better understanding the wolverine's habitat requirements;
- determining the appropriate configuration, size, and appropriate management strategies for protected areas such as parks and wilderness areas, to make sure they play a role as refugia for unexploited wolverine populations; and
- clarifying the scientific classification or taxonomy of various subspecies of wolverines to make sure we are conserving the full diversity of this wild mustelid in Canada and the United States.

7. *Public Information*

The wolverine is a classic example of a species that, even when it exists in healthy wild populations, will never be observed by most people. In this respect, it is probably even more elusive than the cougar. How, then, do we generate an interest in conserving such unseen, but important, creatures of the wilderness?

The answer must be through better public-education programs, particularly those that can make the knowledge and unique personal experiences of people such as Vivian Banci and Maurice Hornocker more broadly available. Vivian's enthusiasm for her work with wolverines in the Yukon and British Columbia is evident when she talks about the importance of education: "It's really important, making sure the public understands how lucky we are with the complement of wildlife that we have and the diversity of habitats. I feel incredibly fortunate to be able to live here and to do what I do. And I'd like other people to feel that as well, to know why it's so special. Then more emphasis can be put into insuring that we keep that specialness, that we don't lose it. I talk to trappers, naturalists, hunters, to newspapers and on the radio — I like it, being where I am, doing what I do, because at least I have a chance to do something. I hope I can make a bit of a difference. If I can get people to listen and to talk, then that's the only objective I have."

Just enough is known about wolverines to give a tantalizing insight into their ecology and their important role as scavengers in natural systems. But not enough is known about them; we can all still want to know more, and therefore make our best efforts to keep them around. The wolverine is a fascinating, yet rare, and seldom-seen predator that is out there right now, wandering over vast areas of remote wilderness. Think of it, roaming effortlessly in a driven search for food through mountain passes, valley bottoms, timbered slopes, marshy bogs, boreal flats, barren lands, deep snows, and arctic ice. Think of it, springing traps and snatching bait, or driving off virtually anything else that wants to share a hard-found source of sustenance.

The wolverine cares little for you and me. But if you are able to appreciate something for its own sake, then what deserves more respect and interest than a wild living creature so entirely capable of living a life independent of people?

8.
A CONSERVATION STRATEGY FOR LARGE CARNIVORES IN CANADA

Unless we as a species learn to control our numbers, our behaviour and

our use of resources in a way that is harmonious with the ecological systems

we live in, over the long term — and this is projecting ahead even

a hundred years — the large carnivores in Canada

do not stand a hope in hell.

WAYNE McCRORY, BEAR BIOLOGIST, BRITISH COLUMBIA

ALTHOUGH EACH OF THE SIX TOP PREDATORS IS SIGNIFICANTLY DIFFERENT, some conservation steps can be taken to benefit all of them. To accomplish this, World Wildlife Fund has outlined a *Conservation Strategy for Large Carnivores in Canada*, made up of seven strategic steps for action.

The Large Carnivore Strategy, as released by WWF for public discussion in November 1990, was based on input from many experts across Canada. We now have the benefit of widespread reaction and some suggested improvements to the initial strategy. On balance, the response has been very positive, with over 100 articles in the popular press; requests for copies of the document from across Canada, the United States, and abroad; extensive support from scientists, experts, and representatives of non-government organizations; and, most important, some clear interest on the part of governments in implementing specific aspects.

A Two-Pronged Approach

Underlying the seven specific conservation steps in the Large Carnivore Strategy are two general approaches to the conservation of top predators. The first approach is what we would call "preservation," which means establishing protected areas and preserving habitat where, as much as possible, our wild hunters are left alone to live their lives unaffected by the activities of people, particularly industrial development. Examples would be wilderness areas within national parks, and certain provincial and territorial parks, ecological reserves, and some wildlife areas. In addition, legislation that totally protects wildlife, such as laws prohibiting any killing, trading, or habitat destruction related to endangered species, might also be considered to be part of the "preservation" approach.

The second approach is what is now categorized under the rather bureaucratic heading "integrated resource management." This approach accepts that wildlife, including top predators, will be affected by the activities of people. Consequently, it emphasizes the need to manage or modify human activities in such a way that the impacts on wildlife are minimized and kept at levels that can be sustained by wildlife populations. Although this approach is often called "wildlife management," notice that it is not wildlife we are managing, but human behavior that affects wildlife. Examples relating to top predators would be: modifying industrial activities such as logging, mining, and road building; changing agricultural practices; properly regulating human killing of wildlife through sport hunting, trapping, poaching, and predator-control programs; controlling the international trade in animal parts; regulating pollution; and undertaking adequate research to make sure all of these are carried out in the interests of wildlife conservation.

Unfortunately, the "preservation" and "integrated resource management" approaches are often seen to be at odds. The "preservation" approach tends to be supported by non-hunters, wilderness advocates, and animal protectionists. They quite rightly point out that there must be some sanctuaries for wildlife where natural systems are still in a wild state, and where people, as well as wildlife, can find true solitude. They see these areas as useful benchmarks against which to measure the changes we humans are making to the rest of the landscape. The "integrated resource management" approach tends to be supported by wildlife managers in government, wildlife users such as hunters and trappers, and representatives from resource industries such as forestry and mining. They quite rightly point out that, if wildlife is to be saved, we must be practical about it, and properly control the human activities that take place on most of the landscape.

Too often, the proponents of these two different approaches have been engaged in a titanic struggle to insure that their particular approach completely dominates conservation policy. But when both sides have good points, it seems to make more sense to bring them together to support common goals. Stephen Herrero made this point nicely in talking about grizzly bear conservation, when he said, "In order for the grizzly bear to survive, it's going to take a lot of different groups working together. It's going to take not only parks and protected areas; it's going to take cooperation from interested forestry and mining companies, and other developers, and so on. All that has to be planned, implemented, and managed with the grizzly bear's survival in mind."

In other words, if we want to have healthy ecosystems and populations of top predators in Canada 100 years from now, we must commit ourselves to a two-pronged approach of (1) preserving natural habitats through the establishment of protected areas with no industrial development, which are large enough to accommodate top predators; and (2) modifying human activities on those lands and waters that are going to be developed and used by people, to insure the long-term conservation of wildlife.

There are strong social, economic, and political pressures lining up behind these measures in Canada today, just as there are pressures mitigating against them. The job of the Large Carnivore Strategy is to call attention to the situation, and to outline what needs to be done for top predators while we still have the option to do it.

OUR GOAL

The general goal of WWF's Conservation Strategy for Large Carnivores can be simply stated: "*To conserve viable, wild populations of large carnivores in Canada.*"

This goal must be pursued at the species, subspecies, and subpopulation levels for polar bears, grizzly bears, black bears, wolves, cougars, and wolverines because each of these species has important, genetically distinct subgroups in Canada. It must be pursued to insure and restore *long-term*, viable populations of these wildlife species. That means taking *immediate* action so that options for the future are not lost. None of this should be seen as being so long term that effective conservation steps cannot, or should not, be taken right *now*. And, finally, this goal should be pursued because large carnivores are worth conserving for their own sake, for their role in representing biological diversity in Canada, and for their appreciation and use by people. Here, we are trying to recognize not just one motivation for conserving top predators, but many. We believe the different reasons discussed in Chapter 1 can and should be mutually reinforcing, not competitive.

To accomplish our goal, the following seven steps need to be undertaken:

1. Determine population conservation goals.
2. Establish Carnivore Conservation Areas.
3. Control killing by humans.
4. Manage impacts on habitat.
5. Broaden public education.
6. Strengthen conservation research.
7. Improve cooperation.

STEP 1:
DETERMINE POPULATION CONSERVATION GOALS

This first step involves assessing current populations of large carnivores as accurately as possible, then determining which populations
should be maintained at certain levels, which should be enhanced,
and which should be re-established through reintroduction or
restoration programs. As a guiding principle, WWF recommends that
under no circumstances should any population of bears, wolves,
cougars, or wolverines be allowed to decline below safe conservation
levels in Canada. That means maintaining at least minimum viable
populations (MVPs) of all these species, at the subspecies and subpopulation levels. Furthermore, we should support the restoration
of extinct or declining wild subpopulations in Canada, where this is
still possible.

The basic principle behind calculating a minimum viable population of a wildlife species is to determine a conservation "bottom
line." If we allow numbers to sink below that bottom line, we run the
unacceptable risk of the species or population declining toward
extinction. Calculating MVPs is a complicated and still-evolving field
of population ecology. Such calculations, though they are the best
guesses anyone can make for now, are fraught with uncertainty and
assumptions that have been made in the absence of definitive scientific evidence. Nevertheless, the work of various researchers has
helped WWF make best guesses at MVPs for those large carnivores for
which some information is known. Our MVP estimates are only for
short-term population viability of top predators, in other words, for
a 50- to 100-year time frame. Current genetic research suggests that,
for long-term population viability, say for 1,000 years, ten times the
number of animals, and therefore ten times the amount of protected habitat, would be needed.

Although these calculations are imprecise, they are the best
possible to date, and they result in some pretty startling figures.
For example, MVP calculations are 393 for grizzly bears, 148 for
wolves, 78 for cougars, and 313 for wolverines. Using known information on the home-range or habitat requirements of these
species, such MVPs translate into very large space requirements.
Examples are in the order of 19,650 to 78,600 square kilometers
(7,600 to 30,400 sq. mi.) for a minimum viable population of

grizzly bears, and between 26,650 and 56,990 square kilometers (10,300 and 22,000 sq. mi.) for wolverines. The areas required for minimum viable populations of wolves and cougars are difficult to calculate because of the great variation in wolf and cougar densities occurring in different regions of Canada.

David Lohnes, reviewing the WWF Large Carnivore Strategy on behalf of the Canadian Parks Service, noted two potential political weaknesses of the MVP approach. First, by providing a convenient "bottom line" for politicians and others facing tough decisions under conflicting pressures, these numbers could be used to go only for the *minimum* size of an area being proposed for protection. Second, to some extent, even the "minimum" gets chipped away by other interests over time, as attested by national park history. This raises the foreboding practical problem of protected areas for top predators shrinking over time, with no opportunity for enlargement. Fred Bunnell, from the University of British Columbia, has succinctly expressed his concerns about MVP calculations: "There's a fearful asymmetry to such estimates. We lose nothing if the estimate is too high; we lose everything if it is too low." And Robert Sopuck, a natural resources policy adviser from the Government of Manitoba, suggests that MVP calculations might best be used to determine guidelines only, which would include a minimum 15 to 20 percent add-on buffer. Then, "even with some chipping away, the add-on buffer would help ensure that the final number chosen did not go below the original MVP."

Clearly, minimum viable population calculations are not perfect. And we should be trying to maintain top predator populations wherever they occur, not just in protected areas. But, if we want to get some reading on what contribution protected areas will make to conserving large carnivores, these calculations can serve as useful guidelines in determining how large and where such areas should be. In addition, if these areas are made large enough to conserve a MVP of the top predators, then we are likely making it big enough to maintain the integrity of many other species and the ecosystem as a whole.

For any species, it is very important to determine the level of genetic distinction or subpopulation that will serve as the target for conservation efforts. For example, we have already seen that it doesn't really make sense to talk about conserving "the wolf" in

Canada because there are many different types or genetic variations of this animal in our country, possibly seventeen subspecies in all. Similarly, there are probably twelve subpopulations of polar bears, at least fourteen "grizzly bear zones," ten subspecies of black bears, four subspecies of cougars, and perhaps two subspecies of wolverines. In the course of preparing the Large Carnivore Strategy, WWF's Scientific Steering Committee "cut the cake" somewhat differently. They identified large carnivore subpopulations according to different major regional habitats or biomes occupied. Wayne McCrory, from British Columbia, points out that, within these larger biome classifications, there are still finer ecological sub-classifications of the landscape. For instance, there are fifty-nine B.C. Parks "regional landscapes" and seventy-three "wildlife ecoregions" in British Columbia alone. These could contribute significantly to the different "ecotypes" of top predators.

What are we do with all of this? At what genetic level should we direct our conservation efforts? This technical question, important as it is, cannot be answered in this book. For now, WWF is simply recognizing and recommending that the conservation of large carnivores in Canada must be directed beyond the species level, to include subspecies, geographically distinct subpopulations, and ecotypes. This recommendation has already been reinforced and supported by various government and non-government agencies. Obviously, then, when determining population conservation goals, our objective must be to conserve some reasonable diversity of the species under consideration. It may be that by pursuing the goal of WWF's Endangered Spaces campaign, which would establish a network of protected areas representing each of the 350 ecological regions of Canada, that the goal of conserving sufficient genetic diversity in large carnivores could also be met.

However, the establishment of these protected areas for our wild hunters should not be used as an excuse to be reckless everywhere else. Otherwise, parks and wilderness areas simply become "islands of green" against a backdrop of degraded wildlife habitat. As Fred Hovey, a grizzly bear researcher, put it: "Although preserves are important, and we need some of them, we also have to be careful that we get the message home that the best pieces of land for the wildlife are the ones humans want as well. By just dwelling on the preserves aspect, do we lose the ability to make effective management strategies in areas outside of preserves?"

When considering Step 1 of the Large Carnivore Strategy, we must also recognize that wildlife populations don't line up neatly along national, provincial, or territorial boundaries. Consequently, setting population conservation targets must be a multi-jurisdictional undertaking. Furthermore, the degree to which these jurisdictions cooperate is a measure of how well a national concern has been addressed. British Columbia and Alberta, for example, have virtually all of Canada's cougars. Therefore, these two jurisdictions must conserve these wild cats on behalf of all Canadians. The point is that population goals must be arrived at cooperatively and on a regional basis, spanning more than one jurisdiction, in keeping with the biology of the species. In this case, if we are to succeed, our political distinctions must be flexible enough to accommodate nature. This may well be *the* major challenge to implementing the *Conservation Strategy for Large Carnivores in Canada.*

STEP 2:

ESTABLISH CARNIVORE CONSERVATION AREAS

Carnivore Conservation Areas (CCAs) are defined here as "areas of sufficient size and managed in such a way to ensure long-term survival for free-ranging, minimum viable populations of large carnivores."

These areas should be established in the context of what Aldo Leopold called "an ecological conscience," and should be part of a broader "land ethic" focusing on ecosystems. CCAs are not places where large carnivores are conserved while they are wiped out everywhere else. Many protected areas in Canada, including parks, wildlife areas, nature reserves, ecological areas, and wilderness zones, are already helping to conserve top-predator populations. However, most were not established for this purpose, so they may be serving such a function almost by accident. Therefore, some national assessment of the adequacy of our existing network of federal, provincial, and territorial protected areas in preserving the genetic diversity of Canada's wild hunters is necessary. It is not the purpose of WWF's Large Carnivore Strategy to propose an entirely new network on top of what we already have. Rather, we are interested in determining what might have to be added, and how existing protected areas could better serve the goal of conserving this important family of animals.

176

In locating CCAs we must also keep in mind the importance of linking them, using natural corridors, and we must also be aware of current and planned human developments such as roads and cities nearby. Paul Joslin offered these observations about the situation in Washington State, where wolves are beginning to work their way down from Canada into the Cascade Mountains: "Although we have one of the largest national parks in the United States, the Olympics, it is extremely unlikely that they will ever reach there on their own because of the major highway, the cities, and a whole series of suburbs in between. In essence, large carnivores require large pieces of land, and, in our traditional way, we've taken and chopped things up into little postage-stamp areas. We need to be thinking of parks and reserves as part of a network that somehow needs to be linked. Those links are every bit as important."

Canadians need to identify specific key areas to protect large carnivores, with a suggested short list for priority action. To get the ball rolling, and keeping our MVP calculations as well as the space required to conserve such populations in mind, five potential CCAs have been identified, based on WWF-supported work through the University of Calgary. Two candidate CCAs are in the Northwest Territories, one is in the Yukon, one is on the Alberta/British Columbia border, and one is in northern British Columbia (see Figure 9). Detailed descriptions of the proposed boundaries and wildlife of these areas are presented in Appendix F. These five areas emerged as obvious candidate CCAs, but they are not meant to be an exhaustive list of all the areas left in Canada that qualify, or as a catalogue of all the areas that would be necessary to conserve minimum viable populations of all top predators in Canada. Also, these areas have been identified primarily on the basis of a biological, not a political or legal analysis. Some of the lands involved span federal, provincial, and territorial jurisdictions, and some of them are still subject to land claims by aboriginal peoples.

These are good and obvious candidate CCAs, but other areas are needed that would provide for differing levels of protection, and for conservation of different subspecies and ecotypes of all the species under consideration. Because of financial constraints, the WWF study focused principally on western Canada; a similar examination urgently needs to be undertaken to identify candidate areas east of Hudson Bay. However, already, additional candidates have been

Figure 9: Proposed Carnivore Conservation Areas

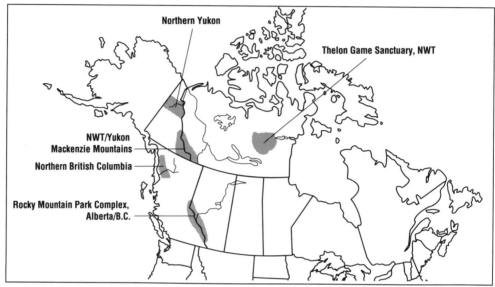

Source: A. Bath, H. Dueck and S. Herrero. Carnivore conservation areas. Draft Report, University of Calgary, 1988.

suggested by interested individuals across Canada.

For coastal British Columbia, Bruce McLellan recommends focusing on the Tweedsmuir Park area. WWF has been working for some time to establish Canada's first grizzly bear reserve in the Khutzeymateen Valley in northwestern coastal British Columbia. Wayne McCrory supports the Khutzeymateen as a CCA candidate, and also proposes a reserve for the Kermode bear on islands off British Columbia. Ric Careless of the B.C. Wilderness Tourism Council, and many others including WWF, need your support to protect the Tatshenshini River Valley in northwestern British Columbia for wolves, grizzly bears, and blue-phase black bears. The "Tat" could be added to the Spatsizi CCA candidate already identified by WWF for that area.

For the southern Rockies, Bruce McLellan has suggested establishing a number of specific wilderness "corridors" to solidify the Rocky Mountain Complex CCA candidate. Susan Hall, a biologist for Mount Revelstoke National Park in British Columbia, organized a 1991 grizzly bear workshop to discuss that area as a specific CCA candidate. And Bob Ream and John Weaver from the University of Montana have emphasized the importance of extending the

Canadian Rocky Mountain carnivore conservation area down into Glacier National Park in the United States.

For the Northern Rockies, Bruce McLellan draws our attention to the Sikanni Chief Area for "white-tails, mule deer, elk, moose, caribou, stone sheep, goats, bison, grizzlies, black bears, wolves, wolverines, and cougar. I don't think there is anything like it outside of Africa!" Pat and Rosemarie Keough urge inclusion of the Ram River and North Nahanni River watersheds in the CCA focused on the Nahanni area of the Yukon/Northwest Territories.

For the tundra region, Kevin Lloyd, Director of Wildlife for the Northwest Territories, has suggested that in addition to the Thelon Game Sanctuary, consideration should be given to establishing a number of small CCA "cores" throughout the barren-grounds, protecting key bear habitat or high-density locations of bears. Then, wilderness corridors where hunting and development activities would be strictly regulated could link the various northern CCAs.

Lu Carbyn from the Canadian Wildlife Service suggests Wood Buffalo National Park in northern Alberta/Northwest Territories, and Riding Mountain National Park in Manitoba for consideration as possible CCAs. Arthur Hoole, Wildlife Director for Manitoba, agrees that Riding Mountain National Park should be considered, for wolves, black bears, and even cougars! He further suggests the proposed Churchill National Park for polar bears and as one area for wolverines.

In the East, John Theberge, of the University of Waterloo, suggests Polar Bear Provincial Park in Ontario — the second-largest park in Canada — and Algonquin Provincial Park in Ontario. In Quebec/Labrador, the Torngat Mountains candidate area for a national park should be evaluated for its potential contribution as a CCA. And in southern New Brunswick, a remote area near Fundy National Park is being supported by Friends of the Eastern Panther, to help protect that elusive endangered species.

This kind of response is encouraging, and indicates that the idea of CCAs is catching on among a wide range of interested professionals across Canada. However, since these proposals all deal with public lands, they need to be supported by all Canadians, not just a select group of technical experts, if things are to happen.

Assuming that some of these candidate areas could be actually protected, then the most difficult and controversial aspect of

managing a CCA is whether or not to permit any killing of large carnivores by people in any part of such an area. The work done for
WWF suggests that there should be large core zones where killing is
not permitted for any purpose. However, the core zones of CCAs
could be surrounded by secondary areas, where carefully controlled hunting *might* be permitted. WWF is further recommending
that there be no predator control, or any other special incentives to
kill predators, within the core *or* in the secondary areas that make
up a CCA. We believe the ecological system encompassed by a CCA
should be as self-regulating as possible, and thereby serve as a
benchmark or example of natural rates of change and fluctuation
of predator/prey numbers. This would also insure that, if hunting
of species such as moose and deer took place in some secondary
zones, it would have to be carried out within natural population
fluctuations, not within manipulated surpluses of game created
through controlling predator numbers.

This concept of zones, beginning with a large core protected
area and moving out to include areas of carefully regulated human
intervention, would allow for a total area of sufficient size to accommodate the ecological needs of large carnivores, without posing
unrealistic demands for strict protection over excessively large areas.
However, it must be emphasized that *the guiding criterion for all management zones — from the most protective to the most permissive — would be
to maintain minimum viable populations of large carnivores.* Therefore,
any activity proposed or permitted in secondary zones that could be
shown to work against this primary interest would be disallowed.
Indeed, the proponents of such activities should have to demonstrate that top predators would *not* be negatively affected if they
were to be allowed.

Most of the candidate CCAs identified in our study are centered
or superimposed on an existing protected area, for example, the
Rocky Mountain Park Complex, the Thelon, the Spatsizi, and the
Nahanni. Although the various parks and reserves in these areas
may not have been originally established with wildlife conservation
in mind, it happens that they have evolved to serve such a purpose.
Now the time has come to recognize this fact and to make the best
of this opportunity.

Cooperative management arrangements in all CCAs must extend
beyond governments and include participation by aboriginal

peoples, guides and outfitters, conservation groups, ranchers, trappers, hunters, industry representatives, and university experts. Peter Clarkson, from the Northwest Territories, offered this comment about cooperative management as the way to go in the future: "We are seeing a lot of cooperative management programs, including working groups with representatives from different backgrounds in society. We're seeing everyone getting together and saying, 'Okay, we have our differences, but let's work out something that's agreeable to everyone!'" The success of the CCA concept will also depend on broad public support of the idea. Therefore, the purpose behind CCAs should be clearly explained in educational material, with the opportunity for citizen feedback on how well the objectives are being met. There is an important, *mutual* education process that must take place, primarily between government agencies that have the legal responsibility for wildlife conservation, and a growing number of Canadians who are expressing great interest in such matters.

Obviously, the exact location and management regime for CCAs remains to be determined in any great detail. Furthermore, the general concept must be of interest to government agencies responsible for implementing it, or it will go nowhere. Therefore, the purpose of this brief outline and the Large Carnivore Strategy is to simply "put the idea out there," to stimulate sufficient discussion to work out the specifics if the idea has merit. Certainly the idea of CCAs is generating encouraging initial support. Ian McTaggart-Cowan wrote, in *Endangered Spaces*: "The new venture would capture imaginations worldwide and put Canada in the forefront. We would be embarking upon a relationship between Canadians of the 21st Century and the impressive species that have been intimately associated with the history and the vision of our land."

STEP 3:
CONTROL KILLING BY HUMANS

Each of the top predators is subject to direct killing by humans through legal recreational and subsistence hunting and commercial trapping. We have stated best guesses at the known levels of kill for each of the species under consideration. However, it is not the purpose of this book to review the principles of setting sustainable harvest levels because these are well known by the management agencies responsible for establishing and enforcing them. By and large, and although mistakes are made, WWF believes efforts are made to do this conscientiously throughout Canada. And this must *always* be the case, because such human killing is a major factor in the overall mortality of the top predators.

For the six top predators featured in this book, estimates of the total number legally killed by people every year in Canada are as follows: 22,000 black bears; 8,000 wolves; 1,000 wolverines; 700 polar bears; 650 grizzly bears; and 250 cougars, for a total of 32,600 animals. To convey an idea of what these numbers mean, if the hides of these animals were laid out side-by-side in the area of an American football field (100 by 53 yards), allowing a 3- by 2-yard area for each animal (some hides would take more space, some would take less), they would occupy thirty-seven full playing fields. This calculation leaves out thousands more animals killed by people illegally through poaching or poisonings, vehicle collisions, and all others caused by humans and not reported. So, in fact, we are talking about many more football fields full, likely a total of at least fifty, every year.

Although this annual kill may be considered sustainable, the sheer volume is staggering. But probably the biggest challenge is to make sure that government wildlife managers maintain their independence to effectively control the legal kill by "calling it like it is." This is because hunting, in particular, has become a very emotional issue on all sides. Those against hunting inevitably argue that the level of kill is too high, or that hunting shouldn't be permitted at all. Those in support of hunting often select information to support their view that everything is okay, or that more animals could be safely killed. Usually, the truth lies somewhere in between. Above all, then, those who set the regulations must find that independent ground and stick to it with professional integrity.

The problem of illegal killing, or poaching, must be better understood and prevented through effective enforcement and education activities. The total size of this illegal kill of top predators is not well known. This uncertainty, when added to the known removal of large carnivores annually from the Canadian wildlife population, should be of serious conservation concern. For example, the 700 polar bears killed through legal hunting every year, excludes unknown or illegal kills. Cougars have been overhunted through legal kills in some specific wildlife-management units, without count-ing other forms of killing. And the illegal killing of grizzly bears, which are being overhunted in many areas already, is probably of even greater conservation concern than that of either polar bears or cougars. It is very difficult to safely control human killing of wildlife species when the size of a potentially important source of mortality, such as poaching, is virtually unknown. This concern applies to all the large carnivores, even to those with more robust reproductive rates, such as wolves, and to those that are still abundant, such as black bears. For bears, the problem may be critical. Fred Bunnell has estimated that in thirty-seven of thirty-nine North American bear populations he analyzed, direct human killing was the major cause of death among adult bears.

Another concern with respect to direct human killing of top predators arises from attempts to deal with problem predators. Some common-sense steps should be taken by livestock owners to minimize conflicts with bears, cougars, and wolves, short of killing the competing predator. Such measures include not allowing live-stock in places where, or at times of the year when, they're likely to be particularly vulnerable to wild predators; proper fencing; proper disposal of carcasses; use of livestock guardian dogs; and even accepting the fact that certain losses are inevitable if one chooses to establish a farm or ranch in prime wildlife habitat. Government compensation programs for livestock losses may also help in this regard. Subsidies to encourage agriculture in wilder-ness settings where it would not otherwise be economically possi-ble should be eliminated.

A final concern is the increased human killing of top predators for profit. As we have all seen in the case of the African rhinoceros and elephant, international trade in animal parts can quickly take a species from a healthy population status to the brink of extinction. Canada is a full party to CITES for this reason (see Appendix D), but we must also be concerned about trade in more abundant species, for example, black bears, which may not find their way onto the CITES Appendices until their numbers decline to dangerously low levels. In fact, all bear species, except the black bear, are now listed in Appendix II, and WWF is recommending that black bears be listed as well. A complete review of Canadian legislation and the adequacy of enforcement is needed immediately in this regard.

STEP 4:
MANAGE IMPACTS ON HABITAT

All of these top predators are subject to manipulation of their natural habitats, which can have a direct impact on their numbers. Such activities include logging, mining, noise (from tankers, aircraft, roads, and all-terrain vehicles), forest spraying (which affects berry crops for food), marine oil spills, and disturbance of specific denning sites. As a Manitoba government reviewer of the Large Carnivore Strategy commented, "In many instances, it is a land-use strategy which is required first, rather than a species-specific strategy."

A growing body of research and practical information is available to wildlife managers, both inside and outside government agencies, to help modify these human activities to lessen their impact on wildlife. It is incumbent upon private companies and public agencies to be aware of this information, and to act voluntarily on it. It is also incumbent upon governments to ensure that such measures are legally required through regulations and legislation such as environmental assessment. These laws to protect wildlife should be encouraged and supported by all Canadians.

Large carnivores are also subjected to indirect habitat impacts from industrial and agricultural activities. Forestry operations in old-growth forests, and livestock grazing in alpine areas, for example, can affect distribution of prey for bears and wolves. Garbage dumps can make "problems" out of wild animals, simply because we put *our* garbage into *their* habitat. Roads make previously inaccessible wilderness backcountry available for increased legal and illegal hunting, or result in the killing of "nuisance" wildlife. Fred Hovey is clearly aware of how serious this problem is becoming: "To get access, people will build bridges, and if they have to, even use winches. I'm not saying close every single road, just shut the roads that go into side valleys and drainages. People now have access onto tops of mountains! One of our black bears was killed by someone who built a skidtrail right up a mountain into an alpine basin! It was amazing. I didn't think they could put a road up there, but they did. And, of course, a black bear was there, foraging on huckleberries, someone shot it, then just left it. That bear would be alive if it weren't for access into every single little drainage!" Recreational developments, such as lodges, hotels, ski hills, and some backcountry hiking facilities, can cause similar problems for wildlife as people invade their natural domain.

All these habitat concerns must be understood, appreciated, and sensitively incorporated into the day-to-day operations of humans and their business when venturing into wilderness areas harboring top predators.

STEP 5:
BROADEN PUBLIC EDUCATION

Reliable information should be demanded by and made available to Canadians regarding the behavior and ecology of large carnivores, especially given the misinformation surrounding these species. Opportunities to view, photograph, and interpret these animals would be welcomed by all of us, residents and non-residents alike, and would also help promote the conservation of our wild hunters. For example, grizzly bears can be viewed while feeding on salmon, polar bears congregate at Churchill before freeze-up, and wolf howling has become a popular activity in Algonquin Provincial Park in Ontario. However, such activities themselves must be controlled to insure that they do not harass wildlife or disturb habitat. Although large carnivores are naturally elusive and difficult to see, good information on identifying and tracking signs of wildlife would help thousands of interested Canadians and tourists feel at least some contact with these unique wildlife species.

As a society, we need to grapple with some major ethical questions regarding allocation of prey to large carnivores, rather than to humans. We need to decide whether or not researchers should use "respectful" management versus "tinkering" by killing animals such as wolves and bears in wild populations for experimental or manipulative purposes. Because other techniques exist to assess the predatory effects of large carnivores on ungulates, and because WWF's Large Carnivore Strategy is built on the inherent worth of large carnivores, we believe these species should not be subjected to experimental control programs. When top predators are marked or radio-collared for research purposes, proper veterinary care in the field must be required, and use of such devices as expandable radio-telemetry collars, or those that will rot off after they have outlived their usefulness, should be encouraged. "Intrusive" science not only can be inhumane, but can contaminate its own results by disturbing wildlife so much that abnormal rather than normal animal behavior is being observed.

Another important ethical question is posed by situations where a particular prey population may be threatened with possible extinction from a specific area. Examples of this are woodland caribou in some areas of British Columbia and Quebec, where wildlife biologists believe that if wolf control is not at least temporarily undertaken, the result will be the loss of the caribou. Do we leave things alone and let the caribou disappear from these areas, even if humans caused the predator/prey imbalance in the first place? Or do we intervene temporarily to "save" the caribou? WWF believes that, if predator-control programs for this purpose are carried out, they must be undertaken in conjunction with other conservation measures, such as halting hunting by humans and controlling other activities affecting the caribou's critical habitat. In this way, responsibility for the long-term survival of the caribou is *shared*, not just foisted off on a natural predator, in this case, the wolf. These are not questions left just for the "experts" to decide, they are ethical concerns where *your* views are needed for direction. As is so often the case, appropriate management practices by the experts in these instances flow from the rest of us having first made up our minds about more fundamental ethical choices.

A final public-information concern focuses on human safety. Canadian jurisdictions such as the Northwest Territories are world leaders in informing us about safety around potentially dangerous large carnivores, particularly grizzly and polar bears. Alberta and British Columbia have similar programs regarding cougars. At the same time that governments encourage and respond to increased interest in these species, it is important that they also advise everyone how to behave safely to minimize the chances of a dangerous encounter. One bad incident, which may only take seconds and might have been avoided, can set wildlife-conservation efforts back by many years.

STEP 6:
STRENGTHEN CONSERVATION RESEARCH

Although we have identified specific research needs for each species, some research needs are shared across the large carnivore groups. Work is needed on the complicated multi-predator/prey systems, upon which so much of Canada's wildlife management is based. In general, more long-term research is necessary to avoid jumping to conclusions about predator or prey population trends based on a "snapshot" in time. Benchmark research on large carnivore populations that are relatively undisturbed by people is extremely useful in understanding human impacts on other wildlife populations. And better systems need to be developed, and funded, for monitoring population levels and causes of death, especially where predator/prey systems are subjected to impacts by people.

All of this research generates the raw data necessary to conserve top predators. Without it, the task is difficult, if not impossible. So, conservation research, too, needs your support.

STEP 7:
IMPROVE COOPERATION

The importance of getting our governments to work together on the conservation of top predators cannot be overemphasized. It also means governments fostering cooperation *within* their borders. For example, Kevin Lloyd, from the Northwest Territories, has noted that, "the establishment of CCAs in the Northwest Territories will occur only with the support and full participation of people in the communities. Consultation and native representation on committees or boards will be necessary for identifying and establishing CCA boundaries and layered buffer zones, as well as later management duties." Lloyd cites many good models for jurisdictional cooperation developed with aboriginal peoples in the Northwest Territories.

This cooperation must extend across national boundaries as well. Bob Andrews, Wildlife Director for the Government of Alberta, for example, has emphasized the importance of working together: "We do see value in jurisdictional cooperation and have participated in similar efforts such as grizzly bear [recovery] in the United States." Harold Dueck, in his WWF-funded study, examined a number of models for interagency cooperation, assessing their strengths and weaknesses, including two from outside Canada. For its own part, WWF has chosen to focus on the Rocky

Mountain ecosystem identified as a candidate CCA, using it as a case-study in the kind of cooperation needed between two provinces, the Canadian Parks Service, and the United States for the effective conservation of top predators.

History has shown that, if deliberate efforts are not made to conserve large carnivores, they are doomed. In many cases, their fates have been sealed *despite* the best efforts of conservationists because those efforts came too late. An alarming number of wildlife experts have already accepted that the long-term destiny for Canada's predators will be extinction, or at best, a few relic populations confined to national parks, as is the case in the United States. But Canada still has the opportunity to do it differently, by not repeating the mistakes made elsewhere. All of us still have the chance to send a clear message to decision-makers in this country. WWF is trying, through this book and through the *Conservation Strategy for Large Carnivores in Canada*, to alert Canadians to this singular opportunity to do things differently. We are indicating what needs to be done, and issuing a warning about what the consequences will be if Canadians choose not to do it. However, we will take little satisfaction in saying "I told you so" fifty years from now.

The *Conservation Strategy for Large Carnivores in Canada* has been developed in consultation with Canada's leading biological experts, and the federal, provincial, and territorial agencies responsible for implementing it. That is not to say that all of them agree with every word, but their input and expertise has been, and will continue to be, honestly sought. WWF has also submitted the strategy for consideration and action at the Federal/Provincial/Territorial Wildlife Conference. This much WWF can undertake. We believe large carnivores are not only worth saving, but that Canada would be profoundly impoverished by their disappearance from our remaining wilderness. However, in the end, it will be up to the Canadian public to make the crucial difference as to whether or not these concerns are taken seriously by decision-makers in our country.

9.
A CONSERVATION STRATEGY FOR LARGE CARNIVORES IN THE UNITED STATES

by John Murray

The conservation of predators is one of the key issues facing people in the United States in the 1990s. Public perceptions of carnivores are undergoing an historic and unprecedented transformation. As a result, there is a rare window of opportunity to set aside new areas for the conservation of these controversial species, as well as to more fully protect and augment existing refuges. In addition, the pressure on critical predator habitat in some areas by commercial mining, timber-cutting, livestock grazing and recreational development is fast reaching a critical point. Increasingly, both public sector policy managers and private sector conservationists are realizing that a comprehensive new approach must be taken if we are to preserve large carnivores for posterity. If we fail, the United States could lose forever these last vestiges of our frontier heritage, just as Germany, Great Britain, and other European countries have lost their brown bears and wolves. Much of what we do, or do not do, in the next ten years will determine the fate of these species in the long term. Conservation efforts need to be directed in five broad areas: land issues, predator restoration, systematic studies, public education, and law enforcement.

Land Issues
Anyone who has visited Yosemite or Yellowstone or Great Smoky Mountains National Park lately knows from the overcrowding that U.S. national parks are insufficient in size and number for the recreational needs of the growing human population, not to mention the habitat requirements of wide-

ranging predators such as grizzly bears, black bears, wolves, and cougars. In the next century we are going to need many more parks, forests, and refuges; the land set aside now is simply not adequate. There are two ways to solve this problem of critical habitat loss. The first is for private groups like the Nature Conservancy to purchase land or obtain conservation easements on private land suitable for wildlife habitats. The second is for the federal government and respective state governments to do the same.

The Nature Conservancy was founded in Great Britain, where there is a paucity of public land and a long tradition of wildlife preservation, particularly bird sanctuaries. The problem facing conservationists in Britain was the same problem now facing conservationists in the United States—a shortage of land for wildlife conservation. The Conservancy approaches the problem either by procuring easements on valuable private tracts or by raising funds to buy the land outright. The organization has been active in the United States for many years and has recently made two important purchases: the 80-square-kilometer (30 sq. mi.) Pine Butte Swamp Grizzly Bear Preserve near Choteau, Montana, and the 1300-square-kilometer (500 sq. mi.) Gray Ranch Nature Preserve near Lordsburg, New Mexico. The Pine Butte Swamp area was established to protect critical spring habitat for the grizzlies along the Rocky Mountain Front in northern Montana. This is the last area in North America where grizzlies still venture out on the high plains. Each spring around a dozen grizzlies come down to the preserve to forage on overwintered berries, fresh spring grass, and succulent roots. The Gray Rand Preserve, located in the bootheel of southwestern New Mexico, protects an island mountain system (Animas Mountains) that once supported grizzlies (the last grizzlies were killed in 1911), Mexican wolves (possibly still present in the area), and jaguars (also reported in recent times in the Animas Mountains). The Gray Ranch also has several large tracts of native grasslands that are in nearly pristine condition; one is approximately 18,500 hectares (44,000 acres) in size. This area would be an ideal location for an experimental restoration of grizzlies, Mexican wolves, and/or jaguars.

In the public sector, funds for conservation are available in the form of royalties levied on oil and gas leases by the federal government. Private mineral development companies have paid enormous sums into an account known as the Land and Water Conservation Fund. Currently, leases generate around $900 million a year, primarily from offshore oil drilling leases on public oil fields. All of these funds are designated solely for conservation purposes and land acquisition. One state where these funds could be of particular value is Florida, where a number of private enclaves threaten to degrade critical cougar habitat. Another is Colorado, where the 32,000-hectare (77,000-acre) Banded Peak Ranch in the south San Juan Mountains possibly contains the last southwestern grizzly bear population. This ranch, which was formerly the northern region of the historic Tierra Amarilla Land Grant, is currently managed largely as a wildlife refuge by conservation-minded owners, the Hughes brothers. Should the ranch one day pass on to other owners who do not share their philosophy, though, the land could be subdivided or developed to the detriment of the bears, as well as to a small endangered population of wolverines and Canadian lynx (both at the southernmost limit of their ranges). Bringing these pristine wilderness lands into public ownership would ensure their protection for endangered predators and for the American people.

Another possible course of action in the public sector—one that has been discussed increasingly in recent years—is to seize critical habitat areas under the "Doctrine of Eminent Domain" principle once used to procure land for national defense purposes (as with a number of ranches in and around the White Sands Missile Range in southern New Mexico in the 1950s). This controversial action would be taken only when owners refuse to sell land at fair market value to the federal government and persist in development activities which threaten animals federally protected under the 1973 Endangered Species Act. The implementation of this doctrine has been discussed most notably with regard to several private land tracts just north of Yellowstone National Park in an area considered to be critical habitat for the threatened grizzly bear population.

One of the major problems facing conservationists with respect to the preservation of predators in the United States is the often bizarre geometric configurations of the national parks in which many of these populations are found. They bear little resemblance to ecosystem boundaries. In his book *Nature First: Keeping Our Wild Places and Wild Creatures Wild* (Roberts Rinehart, 1987), nature writer Thomas McNamee proposed that the federal government create an American system of National Biosphere Reserves, in which nature conservation, as opposed to recreation, would be the primary goal of land management. The respective national parks would be subsumed within these much larger management units. This innovative new system would afford much greater protection for endangered predators. No longer would endangered species managers deal with an often confusing group of Forest Service, National Park Service, and Bureau of Land Management officials; rather, a unified management group would pursue a single course of action with respect to major policy decisions.

Even if the biosphere reserve plan is eventually implemented, conservationists will still have to deal with the problem of fragmented landscapes. This issue becomes particularly important when dealing with isolated populations of predators; for example, the grizzly bears of Yellowstone, the cougars of southern Florida, and the wolverines of Colorado. The problem is that the survival of mammals is inextricably linked to genetic diversity. As a result of inbreeding, a population may become weakened and eventually infertile, which has already taken place in the Florida cougar population. According to conservation biologists Otto Frankel and Michael Soulé, "The basic rule of conservation genetics, based on the experience of animal breeders, is that the maximum tolerable rate of inbreeding is 1 percent. This translates into an effective population size of fifty." To maintain minimum viable populations, biologists are going to have to take one of two steps: either periodically insert new individuals from dislocated populations to maintain genetic viability, or maintain travel corridors between isolated populations so that interbreeding can take place. The long-term survival of isolated predator populations rests on implementing these steps.

Several other innovative ideas have been suggested recently on the subject of critical habitat preservation. One of the most controversial is the so-called "Buffalo Commons" project proposed by several biology professors at Rutgers University. Under their plan, lands in the northern plains presently under cultivation or being grazed by domestic livestock would be allowed to return to a native state. Eventually, sizeable herds of buffalo would be returned to this vast area, along with wolves and bears. Those who support the "Buffalo Commons" concept contend that agricultural practices have had a catastrophic effect on the fragile grasslands, and that the most viable economic and environmental option is to return this area to its historic state. Whatever happens to the "Buffalo Commons" proposal, and others like it, it is clear that the issue of critical habitat for predators will be center stage in the 1990s; the American people are going to have to make some difficult decisions about their priorities with respect to these endangered animals, for the predators can only persist in the next century through our generosity and restraint in this century.

Predator Restoration
Predator restoration is emerging as one of the central themes of U.S. conservation in the 1990s. Two factors are at the heart of this historic movement.

First, the livestock industry, which was responsible for the eradication of grizzly bears, black bears, mountain lions, gray wolves, Mexican wolves, and jaguars throughout the American West, is in a state of severe decline. Feedlots are replacing cattle ranches, consumers are choosing poultry and fish over red meat, federal land managers are increasing grazing fees, woolgrowers are being challenged by foreign producers, and inactive grazing allotments are being purchased by the states for wildlife use or simply retired altogether. The net result is that the cattle ranchers and woolgrowers are no longer the powerful hegemony they once were, virtually controlling federal agencies such as the Bureau of Land Management, the Forest Service, and the Fish and Wildlife Service (formerly the notorious Biological Survey) and dictating management policy.

Second, the "predator prejudice" that for so many centuries treated all predators as agents of destruction to be persecuted and destroyed, has been replaced by a new, more enlightened viewpoint. Increasingly, people understand the inherent value of predators and seek ways of accommodating the demands of their unique lifestyles. This reversal of public opinion is best reflected in the restoration of red wolves in a number of locations throughout the Old South; such a program would have been unthinkable as recently as the 1970s.

There are several places where grizzly bears could be reintroduced in the 1990s. At least one of these—the San Juan Mountains of southwestern Colorado—has been officially included in the 1991 Grizzly Bear Recovery Plan published by the U.S. Fish and Wildlife Service. Attorneys for the agency have determined that under the 1973 Endangered Species Act the federal government is required to protect, and if necessary augment, all grizzly bear populations that were in existence in 1975. Because a relict population possibly still exists in the San Juans, the Fish and Wildlife Service now plans to undertake preliminary surveys and regional studies in the 1990s. These mountains contain some of the finest bear habitat in North America and over 2,400 square kilometers (900 sq. mi.) of wilderness have been designated in the area. Bear habitat there is similar to that found in the northern Rockies, where minimum habitat studies have determined that approximately 2,500 square kilometers (965 sq. mi.) are necessary to support a minimum viable population of grizzly bears (70 to 90 bears). Managers have a number of new scientific techniques at their disposal which would make restoration easier. For example, a technique known as "interspecific cross-fostering" could be employed; this process involves placing newly born grizzly cubs in the natal dens of lactating black bear sows. Highly detailed satellite photographs could help with habitat evaluation and management decisions. Released grizzlies could be radio-collared and radio-mapped. The timing of the releases could be planned, as it is in Denali, Glacier, and Yellowstone, in order to avoid conflicts. The same safeguards used in these other grizzly areas could help in public education, as well, to ensure people camp and hike safely.

Three other areas have been considered for grizzly restoration. The first is the combined Blue Range Primitive Area/Gila Wilderness/Aldo Leopold Wilderness ecosystem in southwestern New Mexico and southeastern Arizona. All totaled, these designated wild areas in the Mogollon Mountains and Black Range preserve well over 2,600 square kilometers (1,000 sq. mi.) of pristine bear habitat. An added advantage is that most of the grazing allotments in the Gila, which historically supported a large number of cattle and sheep, have been retired or purchased by the state for wildlife use. The Gila, often referred to as the "Yellowstone" of the Southwest, was formed as the world's first designated wilderness area in 1924, largely as a result of the pioneering efforts of Aldo Leopold, who later argued for wolf and grizzly protection in his seminal essays "Thinking Like a Mountain" and "Escudilla." The second area is New Mexico's Gray Ranch Preserve, which offers over 1,300 square kilometers (500 sq. mi.) of former grizzly habitat for an experimental release. The third possibility is the Sierra Nevada Mountains of California. From Yosemite National Park south to Sequoia National Park, there are over 5,800 square kilometers (3,600 square miles) of contiguous parkland and wilderness area. A further motivation is the fact that California calls itself the Golden Bear state, and the grizzly is featured on the California state flag. As in the Blue Range/Gila/Aldo Leopold ecosystem and the Gray Ranch ecosystem, black bears currently exist in good numbers in the Sierra Nevada region, attesting to the quality of these areas as bear habitat.

There are at least several areas in the American West that could be considered for wolverine restoration programs: the Sierra Nevada Mountains of California, the High Uintas area in northern Utah, the Flattops Wilderness Area in Colorado, Rocky Mountain National Park in Colorado, and the San Juan Mountains (Weminuche Wilderness and South San Juan Wilderness) in Colorado. All of these regions offer the essential components of wolverine habitat: a wilderness location, a high mountain ecosystem, and an adequate prey base. In *Wildlife in Peril: The Endangered Mammals of Colorado* (Roberts Rinehart, 1987), I recommended that wolverines be restored either to Rocky Mountain National Park or to the San Juan ecosystem. One of the clear advantages of Rocky Mountain National Park is that there is no trapping, hunting, or grazing in the park, all sources of historic conflict with wolverines. Additionally, one of the objectives of the National Park Service management policy is to preserve, and if necessary restore, pre-Columbian conditions in the parks. The restoration of the wolverine would be consistent with the wilderness objectives set forth in the now-famous 1963 Leopold Report. It is possible that wolverines occur in all three of the Colorado locations, as well as in central California, so these restoration projects, like the Colorado grizzly one, could technically be considered augmentation projects.

The primary restoration issue, with respect to the cougar in the United States, is the situation facing the Florida cougar. At the center of the controversy regarding this predator is the captive breeding program. Some biologists, for example, have suggested that the Florida cougar be cross-bred with cougars from healthier populations. At a later time, through back-breeding, a stronger genetic version of the Florida cougar could be artificially produced and released in the wild to augment the current endangered population. Such a complicated breeding program was undertaken for the European wild horse earlier this century. As a result, a number of these once-vanished wild horses are raised in special enclosures in eastern European countries such as Poland. In the 1980s, a back-breeding program was suggested for the black-footed ferret, a small weasel that was at that time one of the most critically endangered mammals in the world (currently there are over 200 in the U.S. Fish and Wildlife Service's captive breeding program). Whatever the case, there is little doubt that the Florida cougar faces a number of years, possibly decades, of management-intensive recovery programs aimed at replenishing wild populations with captive-bred animals and, possibly, at restoring cougars to other areas of former habitat in southern Florida.

The only other restoration area that has been seriously considered for the cougar is Great Smoky Mountains National Park, which straddles the Appalachian Mountains on the North Carolina/Tennessee border. Through the years, a number of cougar sightings have been reported in this large wild region, and one was killed in eastern Tennessee in 1971. If scientific authorities, public policy managers, and local residents can reach a consensus on cougar restoration, a recovery plan should be developed for the species. The cougar could help control the large numbers of European wild boars (not a native species), which destroy native wildflowers throughout the park. The fact that red wolves have been released in this region gives managers hope that one day the cougar will return as well.

From time to time various federal and state biologists and freelance nature writers have suggested that the jaguar be restored to at least one location in its former habitat in the American Southwest. At one time this wild cat was found in south Texas along the Rio Grande, in southwestern New Mexico south of and including the Mogollon Mountains, and in southeastern Arizona through the Chiricahua Mountains and up the major watercourses. If public interest warrants, and scientific authorities concur, jaguars could be eventually returned to one of three areas. Probably the most likely location is Big Bend National Park in southwestern Texas along the Rio Grande. This park, which protects over 700,000 acres, would provide an ideal location because there would be no conflicts with hunting, trapping, or livestock grazing, all sources of historic conflict with the species. The second area which has been mentioned is the Gila Wilderness of southwestern New Mexico, which, like Big Bend, provides a large, rugged, and little-visited wild area with a suitable prey base of mule deer, white-tailed deer, javelina, elk, and antelope.

The restoration of red wolves, Mexican wolves, and gray wolves is much more a reality than jaguar or eastern cougar restoration. One of the factors that positively influenced proposals for wolf restoration in the 1980s was the fact that a considerable amount is known about wolf ecology. In order to eradicate the species, the U.S. Biological Survey animal damage control agents had to fully understand every aspect of the predator's natural history. Ironically, this information actually helped to prove the argument that wolves are not as destructive to livestock as previously thought.

One area in the United States that was intensely studied was northern Minnesota. In 1980, it was estimated that there were about twelve hundred wolves and 300,000 sheep and cattle on 12,000 farms in the area. From 1979 through 1981, the greatest single-year losses in northern Minnesota were 30 cattle and 110 sheep in 1981. About 10 percent of the annual complaints involved coyotes. Only a few farms and grazing leases sustained more than one wolf depredation during any one grazing season. Often, only a single farmer sustained any serious losses in an entire year. In 1977, for example, one sheep farm received 65 percent of the total compensation paid out by Minnesota for that year. In 1978, a single cattle ranch received 42 percent, and the same ranch was paid 51 percent of the total the following year. Minnesota paid up to $400 per animal on verified wolf kills or injuries during this period. From 1977 through 1980, the state paid farmers a total of $72,381.82 on eighty-six of ninety-three claims.

These facts—bearing in mind that we are talking about twelve hundred wolves—actually helped proponents of wolf restoration, whether the red wolf in the Old South, the gray wolf in Yellowstone, or the Mexican wolf in the Southwest, to make a strong case for restoring these valuable predators in their respective regions of the country. Historically, the strongest argument against reintroducing the species has been that wolves destroy livestock to such an extent that they cannot be tolerated. What the information from the various Minnesota studies clearly shows, however, is that the extent of this damage has been greatly exaggerated by opponents of the predator in the past. The wolf may occasionally prey on domestic livestock, but much prefers its wild prey, and may actually avoid domestic livestock in some situations. The success of the ongoing red wolf reintroductions in the South, and the increasing likelihood of the gray wolf restoration in Yellowstone and the Mexican wolf restoration in New Mexico, indicate how far we have come in a very short time. Just one generation ago, such projects would have been laughed out of the room, but today wolves are returning—either through transplanting or through natural colonization—to several areas in the country.

Public Education
The dissemination of accurate information and pro-predator viewpoints, particularly among the younger generations, is essential to preserving the future of these species in the United States. The goal of educating the public about the importance of carnivores in the wild can be accomplished in several ways.

First, and probably most important, is the indispensable role played by public television, especially on the Public Broadcasting System, in presenting nature programs and wildlife documentaries that demonstrate the integral role played by predators in dynamic natural systems. These programs are particularly effective with younger viewers, who, in a short period of ten or fifteen years, will be the emerging leaders of the next century. In an era of fiscal conservatism and growing pressure to reduce federal funding for such programs, it is imperative that elected officials be made aware of the fact that these programs are appreciated by their respective constituencies. If anything, additional funding should be made available for public television, because commercial television is diverting more and more funds from "special projects" involving wildlife to what are perceived as more lucrative subjects.

Second, the federal government should fund a new program entitled "Naturalists in the Schools," just as it currently funds the "Poets in the Schools," "Musicians in the Schools," and "Artists in the Schools" programs through the National Endowment for the Arts. These roving naturalists could provide lecture-demonstrations in elementary and junior high school classes that would address environmental issues of interest or concern in the various regions of the country. Again, the importance of reaching children cannot be overestimated; here is where far-reaching changes in orientation and consciousness can be achieved with minimal effort and expenditure of funds.

Third, more periodical articles and books need to be written on predators in order to better educate adult audiences as to the value of these species to the country. Writing competitions are one of the ways that writing about predators could be encouraged. A prestigious award, with a relatively small cash prize and/or publication arrangement, could attract a large number of quality entries, all on the theme of predators. These essays or books could then be widely distributed in the major retail chains and regional bookstores, and reviews and excerpted in the more popular environmental periodicals. This would annually focus a considerable amount of public attention on predators.

Fourth, although such groups as Defenders of Wildlife have taken on projects with respect to carnivores, more organizations are needed to lobby specifically for predators. For example, in 1987, in *Wildlife in Peril: The Endangered Mammals of Colorado* (Roberts Rinehart), I said it would probably be necessary for a Colorado Grizzly Bear Society to be formed. This organization would place public education materials into the schools to dispel some of the misconceptions about the grizzly; would lobby the National Forest Service, U.S. Fish and Wildlife Service, and Colorado Division of Wildlife for grizzly preservation and restoration; and would raise funds. In 1990, the U.S. Fish and Wildlife Service added Colorado to its list of grizzly ecosystems under study, and in 1991 a "Citizens Committee for the Colorado Grizzly" was formed by Tony Povolitis and Dennis Sizemore in Colorado. This group is already pressuring state officials to close the San Juan grizzly ecosystem to hunting and livestock grazing in some areas, and is pressing federal officials for increased protection. This is just the sort of local organization that will be needed in order to protect predators in the decades to come.

Traditionally, both natural history museums and zoological parks provided most of the education *vis-á-vis* predators and their complex roles in nature. The demands of our times are such, though, that museums and zoos are no longer sufficient as the sole conduits for information. Only by aggressively educating the public—especially the young—can the American public hope to preserve predators in the crowded centuries that follow our own.

Systematic Studies
For virtually all of the large carnivore species, additional systematic studies are needed to help delineate habitat requirements and answer questions of natural history. Long-term studies are needed, for example, for both the red wolf and the Mexican wolf in the wild; little is known of their ecology in the diverse environments they have historically occupied in the United States. Gray wolf restoration in Yellowstone National Park will provide an unprecedented opportunity to study the animal as it interacts with native prey populations.

With respect to black bears, long-term studies are needed to more accurately describe the social structure of wild populations, as well as to acquire more comprehensive natality and mortality rates. Biologist Tom Beck's seven-year study of black bears on Black Mesa in Gunisson National Park in southwestern Colorado illustrates how little is really known about wild black bear populations; his data contradicted much of what was believed about the species in the southern Rockies. The primitive censusing techniques now employed are also in need of refinement. Most importantly, with regard to both black bears and grizzly bears, there is a need to better manage recreational visitors in national parks and wilderness areas. The grizzly has been intensively studied in a number of locations in the lower forty-eight states and Alaska. Although there is little point in further harassing the animals through radio-collaring and radio-mapping at these locations, there is a need to better understand grizzly habitat use through satellite imaging techniques, a process pioneered by John Craighead in the late 1970s and early 1980s in the Lincoln-Scapegoat Wilderness in northern Montana and in the Gates of the Arctic National Park in northern Alaska.

Much work remains to be done on the wolverine, as little is known of the ecology and wild behavior of this reclusive carnivore. As with wolves, wolverine restoration programs would provide an excellent chance to radio-instrument and further study the ecology of a little-understood predator. One of the problems with wolverines, however, which is not the case with cougars, wolves, or grizzlies, is that the animals frequently lose their neck collars because of the small size of their heads. Biologists radio-instrumenting river otters have often inserted the instrument package beneath the skin; it is to be hoped a less instrusive and risky method can be found for the wolverine. In the future, the radio packages may be able to be further miniaturized as new breakthroughs in technology—particularly in battery size—are made.

With respect to the cougar, scientists still need to undertake a systematic study of taxonomy to determine what genetically separates the various subspecies found in the contiguous United States. Such data could be important in the Florida cougar captive breeding program. Additionally, more research should be undertaken on the subject of livestock predation by cougars. These projects could accurately determine the real extent of damage caused by this predator. Finally, any reintroduction of jaguars to the Southwest could focus international scientific attention on both the region, which is beset by major environmental problems, and the lifestyle of this little-studied carnivore.

Law Enforcement
Anyone who has lived in rural America very long knows that poaching is a serious problem. This illegal activity involves all types of wildlife: big-game species, small-game species, non-game species, threatened animals, and endangered animals. Some trophy poachers routinely conduct their operations in na-

tional parks, where all forms of hunting and trapping are illegal. This is particularly the case in Alaska, where the national parks and preserves are so large—Denali National Park is larger than the state of Massachusetts—and funding levels for enforcement are so small that it is impossible for poachers to be apprehended on a regular basis.

One enforcement issue that is especially troublesome is that of black bear and grizzly bear poaching. Because the bear gall bladder is prized in Asian cultures for use in medicines, prices for this organ on the black-market have recently exceeded $3,000. Bear paws, a traditional delicacy, fetch up to $500 each in Korea, and bear cubs are sometimes illegally exported and raised in Asian countries, where they are sold for up to $5,000 and later killed for their body parts. Because fur-trapping is no longer economical, with the European Common Market ban on leg-hold trap fur and the growing consumer distaste for wild fur, the economies of small communities in northern Canada and Alaska are in desperate need of an alternative supply of cash. Illegal poaching of bears has unfortunately become the answer in some cases.

There are several ways to combat the illegal killing of predators. First, substantial rewards need to be offered for people reporting information that leads to the arrest and conviction of poachers. These rewards can be offered both by private organizations, such as the Audubon Society, which has offered awards in recent years, and by local, state, and federal law enforcement or wildlife agencies. Second, additional funding needs to be made available at both the federal and the state level to hire more conservation officers to patrol the vast wilderness areas in which predators are found. Third, law enforcement agencies need to infiltrate these often-international organizations and learn as much as they can so that indictments can be obtained that will hold up in court.

Sadly, the problem of poaching will probably grow in the next few decades as predators such as the black bear and grizzly bear become more scarce and demand for their body parts increases. The United States could find itself in the same position with respect to the grizzly, for example, as Kenya and Tanzania have found themselves in with regard to the rhinoceros, which is also valued for its body parts in the Far East and which has also been decimated by poachers. It is to be hoped, however, that alert and aggressive law enforcement agencies can control this problem before it becomes a greater crisis.

In summary, through improved land use involving more and larger protected areas, through efforts to restore predators to their former range, through systematic field studies to gather more scientific information that will be useful for conservation, through improved public education focussing on the young, and through strengthened law enforcement, we still have the chance to maintain viable populations of top predators in the United States. What is needed is the public support and political will to take these steps while we still can.

CONTACTS

The following list is provided as a starting point to assist you in finding out more from your government wildlife agencies as well as from your local non-governmental organizations. This list should not be interpreted as comprehensive; many other groups and agencies work in behalf of large predators. Check with your local government branch offices, as well as with local nature centers and state or national parks for a more complete listing of other wilderness and environmental contacts in your area.

FEDERAL GOVERNMENT

To contact your United States Senator write:

THE HONORABLE _____
United States Senate
Washington, D.C. 20510

To contact your United States Representative write:

THE HONORABLE _____
United States House of Representatives
Washington, D.C. 20515

Chief Administrator
Environmental Protection Agency
201 M Street, S.W.
Washington, D.C. 20460

President of the United States
The White House
1600 Pennsylvania Avenue
Washington, D.C. 20500

Secretary of Agriculture
U.S. Department of Agriculture
The Mall, 12th and 14th Streets
Washington, D.C. 20250

Secretary of the Interior
U.S. Department of the Interior
C Street between 18th and 19th
Streets, N.W.
Washington, D.C. 20240

U.S. Fish and Wildlife Service
Endangered Species
1745 W. 1700 Street
Salt Lake City, UT 84116

U.S. Fish and Wildlife Service
Federal Center
P.O. Box 254846
Denver, CO 80225

NON-GOVERNMENT
ORGANIZATIONS AND
OTHER CONTACTS

Alaska Natural History
Association
Denali National Park, AK
99755-0009

Alaska Wildlife Alliance
Box 190953
Anchorage, AK 99519

American Forestry Association
1516 P Street, N.W.
Washington, D.C. 20005

American Society of Zoologists
Box 2739
California Lutheran College
Thousand Oaks, CA 91360

American Ecological Research
Institute
432 Burr Oak Drive
Kent, Ohio 44240

American Endangered Species
Foundation, Inc.
1988 Damascus
Black Hawk, CO 80422

American Parks and Wildlands
P.O. Box 97
Big Sky, MT 59716

Convention on International
Trade in Endangered Species
(CITES)
Canadian Wildlife Service
Ottawa, Ontario K1A OH3

Defenders of Wildlife
1244 19th Street
Washington, D.C. 20036

Earth First!
18848 S.E. 269th Street
Kent, WA 98042

Friends of the Earth
530 7th Street, S.E.
Washington, D.C. 20003

Greater Ecosystem Alliance
P.O. Box 2813
Bellingham, WA 98227

Green Wolf
205 Cazneau Avenue
Sausalito, CA 94965

Hornocker Wildlife Research
Institute
P.O. Box 3426 University
Station
Moscow, ID 83843

CONTACTS

International Fund for Animal
Welfare (IFAW)
P.O. Box 193
Yarmouth Port, MA 02675

International Wildlife
Coalition
c/o NBS/Concept 1, Inc.
341 Albany Street
New York, NY 10280

International Wolf Center
1900 E. Camp Street
Ely, MN 55731

National Audubon Society
950 Third Avenue
New York, NY 10022

National Parks and
Conservation Association
(NPCA)
1015 31st Street, N.W.
Washington, DC 20007

National Wildlife Federation
1412 19th Street, N.W.
Washington, D.C. 20036

Nature Conservancy
1800 North Kent Street
Arlington, VA 22209

North American Wilderness
Recovery Project
P.O. Box 5365
Tucson, AZ 85745

North American Wildlife
Foundation
1266 W. Northwest Hwy,
Suite 806
Palatine, IL 60067

Sierra Club
100 Bush Street, 13th Floor
San Francisco, CA 94104

Society of American Foresters
5400 Grosvenor Lane
Bethesda, MD 20814

Society for Conservation
Biology
c/o 52 Beacon Street
Boston, MA 01208

Wild Earth
P.O. Box 492
Canton, NY 13617

Wilderness Society
1400 Eye Street, N.W.
Washington, D.C. 20005

Wildlife Education Project for a
Living Future
Rte 2, Box 225A
Bovey, MN 55709

Wildlife Management Institute
1101 14th St. N.W., Suite 725
Washington, DC 20005

The Wildlife Society
5410 Grosvenor Lane
Bethesda, MD 20814

CONTACTS

WOLF!
P.O. Box 29
Lafayette, IN 47902

World Wildlife Fund U.S.
1250 24th Street, N.W.
Washington, DC 20037

APPENDIX A

INTERNATIONAL UNION FOR THE CONSERVATION OF NATURE AND NATURAL RESOURCES (IUCN) MANIFESTO ON WOLF CONSERVATION

This Manifesto comprising a Declaration of Principles for Wolf Conservation and recommended Guidelines for Wolf Conservation was adopted by the IUCN/SSC Wolf Specialist Group at its meeting in Stockholm, Sweden, in September 1973 and has been endorsed by the Survival Service Commission and the Executive Board.

The Stockholm meeting was attended by official delegates and observers from 12 countries having important wolf populations. It was the first international meeting on the wolf.

DECLARATION OF PRINCIPLES FOR WOLF CONSERVATION

1. Wolves, like all other wildlife, have a right to exist in a wild state. This right is in no way related to their known value to mankind. Instead, it derives from the right of all living creatures to co-exist with man as part of the natural ecosystems.
2. The wolf pack is a highly developed and unique social organization. The wolf is one of the most adaptable and important mammalian predators. It has one of the widest natural geographical distributions of any mammal. It has been, and in some cases still is, the most important predator of big-game animals in the northern hemisphere. In this role, it has undoubtedly played an important part in the evolution of such species and, in particular, of those characteristics which have made many of them desirable game animals.
3. It is recognized that wolf populations have differentiated into subspecies which are genetically adapted to particular environments. It is of first importance that these local populations be maintained in their natural environments in a wild state. Maintenance of genetic purity of locally adapted races is a responsibility of agencies which

plan to reintroduce wolves into the wild as well as zoological gardens that may prove a source for such reintroductions.

4. Throughout recorded history man has regarded the wolf as undesirable and has sought to exterminate it. In more than half of the countries of the world where the wolf existed, man has either succeeded, or is on the verge of succeeding, in exterminating the wolf.

5. This harsh judgement on the wolf has been based first, on fear of the wolf as a predator of man and second, on hatred because of its predation on domestic livestock and on large wild animals. Historical perspectives suggest that to a considerable extent the first fear has been based on myth rather than on fact. It is now evident that the wolf can no longer be considered a serious threat to man. It is true, however, that the wolf has been, and in some cases still is, a predator of some consequence on domestic livestock and wildlife.

6. The response of man, as reflected by the actions of individuals and governments, has been to try to exterminate the wolf. This is an unfortunate situation because the possibility now exists for the development of management programmes which would mitigate serious problems, while at the same time permitting the wolf to live in many areas of the world where its presence would be acceptable.

7. Where wolf control measures are necessary, they should be imposed under strict scientific management and the methods used must be selective, highly discriminatory, of limited time duration and have minimum side-effects on other animals in the ecosystem.

8. The effect of major alterations of the environment through economic development may have some serious consequences for the survival of wolves and their prey species in areas where wolves now exist. Recognition of the importance and status of wolves should be taken into account by legislation and in planning for the future of any region.

9. Scientific knowledge of the role of the wolf in ecosystems is inadequate in most countries in which the wolf still exists. Management should be established only on a firm scientific basis, having regard for international, national and regional situations. However, existing knowledge is at least adequate to develop preliminary programmes to conserve and manage the wolf throughout its range.

10. The maintenance of wolves in some areas may require that society at large bear the cost e.g. by giving compensation for the loss of domestic stock; conversely there are areas having high agricultural value where it is not desirable to maintain wolves and where their introduction would not be feasible.

11. In some areas there has been a marked change in public attitudes towards the wolf. This change in attitudes has influenced governments

to revise and even to eliminate archaic laws. There is a continuing need to inform the public about the place of the wolf in nature.

12. Socio-economic, ecological and political factors must be considered and resolved prior to reintroduction of the wolf into biologically suitable areas from which it has been extirpated.

GUIDELINES ON WOLF CONSERVATION

The following guidelines are recommended for action on wolf conservation:

A. *General*

1. Where wolves are endangered regionally, nationally or internationally, full protection should be accorded to the surviving population. (Such endangered status is signalled by inclusion in the Red Data Book or by a declaration of the Government concerned.)

2. Each country should define areas suitable for the existence of wolves and enact suitable legislation to perpetuate existing wolf populations or to facilitate reintroduction. These areas would include zones in which wolves would be given full legal protection, e.g. as in national parks, reserves or special conservation areas, and additionally zones within which wolf populations would be regulated according to ecological principles to minimize conflicts with other forms of land use.

3. Sound ecological conditions for wolves should be restored in such areas through the rebuilding of suitable habitats and the reintroduction of large herbivores.

4. In specifically designated wolf conservation areas, extensive economic development likely to be detrimental to the wolf and its habitat should be excluded.

5. In wolf management programmes, poisons, bounty systems and sport hunting using mechanized vehicles should be prohibited.

6. Consideration should be given to the payment of compensation for damage caused by wolves.

7. Legislation should be enacted in every country to require the registration of each wolf killed.

B. *Education*

A dynamic educational campaign should be promoted to obtain the support of all sectors of the populations through a better understanding of the values of wolves and the significance of their rational management. In particular the following actions are advocated:

a) press and broadcast campaigns;

b) publication and wide distribution of information and educational material; and

c) promotion of exhibitions, demonstrations, and relevant extension techniques.

C. Tourism

Where appropriate, general public interest in wolf conservation should be stimulated by promoting wolf-related tourist activities. (Canada already has such activities in some of its national and provincial parks.)

D. Research

Research on wolves should be intensified, with particular reference to:

a) surveys on status and distribution of wolf populations;

b) studies of feeding habits, including especially interactions of wolves with game animals and livestock;

c) investigations into social structure, population dynamics, general behaviour and ecology of wolves;

d) taxonomic work, including studies of possible hybridization with other canids;

e) research into the methods of reintroduction of wolves and/or their natural prey; and

f) studies into human attitudes about wolves and on economic effects of wolves.

E. International Cooperation

A programme of international cooperation should be planned to include:

a) periodical official meetings of the countries concerned for the joint planning of programmes, study of legislation, and exchange of experiences;

b) a rapid exchange of publications and other research information including new techniques and equipment;

c) loaning or exchanging of personnel between countries to help carry out research activities; and

d) joint conservation programmes in frontier areas where wolves are endangered.

ACCEPTED CHANGES IN THE MANIFESTO ON WOLF CONSERVATION
AND GUIDELINES ON WOLF CONSERVATION:

Item 7 of the Manifesto

It is required that occasionally there may be a scientifically established need to reduce non-endangered wolf populations; further it may become scientifically established that in certain endangered wolf populations specific individuals must be removed by appropriate conservation authorities for the benefit of the wolf populations. Conflict with man sometimes occurs from undue economic competition or from imbalance predator-prey ratios adversely affecting prey species and/or the wolf itself. In such cases, temporary reduction of wolf populations may become necessary, but reduction measures should be imposed under strict scientific management. The methods must be selective, specific to the problem, highly discriminatory, and have minimal adverse side effects on the ecosystem. Alternative ecosystem management, including alteration of human activities and attitudes and non-lethal methods of wolf management, should be fully considered before lethal wolf reduction is employed. The goal of wolf management programs must be to restore and maintain a healthy balance in all components of the ecosystem. Wolf reduction should never result in the permanent extirpation of the species from any portion of its natural range.

Item 11 of the Manifesto

In some cases there has been a marked change in public attitudes towards the wolf. This change in attitudes has influenced governments to revise and even to eliminate archaic laws. It is recognized that education to establish a realistic picture of the wolf and its role in nature is most essential to wolf survival. Education programs, however, must be factual and accurate.

Item B of Guidelines

B. Education

A dynamic educational campaign should be promoted to obtain the support of all sectors of the population through a better understanding of the values of wolves and the significance of their rational management. Public information should be coordinated and should be implemented with the help of professionals. Specific tools and approaches should be designed for different cultural and social settings.

COMMITTEE ON THE STATUS OF ENDANGERED WILDLIFE IN CANADA
(COSEWIC)
DEFINITIONS
(as of 1990)

COSEWIC: A committee of representatives from federal, provincial, territorial, and non-government agencies that officially assigns national status to species at risk in Canada.

SPECIES: Any species, subspecies, or geographically separate population.

VULNERABLE SPECIES: Any indigenous species of fauna or flora that is particularly at risk because of low or declining numbers, occurrence at the fringe of its range or in restricted areas, or for some other reason, but is not a threatened species.

THREATENED SPECIES: Any indigenous species of fauna or flora that is likely to become endangered in Canada if the factors affecting its vulnerability do not become reversed.

ENDANGERED SPECIES: Any indigenous species of fauna or flora that is threatened with imminent extinction or extirpation throughout all or a significant portion of its Canadian range.

EXTIRPATED SPECIES: Any indigenous species of fauna or flora no longer known to exist in the wild in Canada but occurring elsewhere.

EXTINCT SPECIES: Any indigenous species of fauna or flora formerly indigenous to Canada but no longer known to exist anywhere.

APPENDIX C

INTERNATIONAL AGREEMENT ON THE CONSERVATION OF POLAR BEARS AND THEIR HABITAT

THE GOVERNMENTS of Canada, Denmark, Norway, the Union of Soviet Socialist Republics, and the United States of America,

RECOGNIZING the special responsibilities and special interest of the States of the Arctic Region in relation to the protection of the fauna and flora of the Arctic Region;

RECOGNIZING that the polar bear is a significant resource of the Arctic Region which requires additional protection;

HAVING DECIDED that such protection should be achieved through coordinated nation measures taken by the States of the Arctic Region;

DESIRING to take immediate action to bring further conservation and management measures into effect;

HAVE AGREED AS FOLLOWS:

ARTICLE I

1. The taking of polar bears shall be prohibited except as provided in Article III.
2. For the purpose of this Agreement, the term "taking" includes hunting, killing and capturing.

ARTICLE II

Each Contracting Party shall take appropriate action to protect the ecosystems of which polar bears are a part, with special attention to habitat components such as denning and feeding sites and migration patterns, and shall manage polar bear populations in accordance with sound conservation practices based on the best available scientific data.

ARTICLE III

1. Subject to the provisions of Articles II and IV, any Contracting Party may allow the taking of polar bears when such taking is carried out:

a) for bona fide scientific purposes; or

b) by that Party for conservation purposes; or

c) to prevent serious disturbance of the management of other living resources, subject to forfeiture to that Party of the skins and other items of value resulting from such taking; or

d) by local people using traditional methods in the exercise of their traditional rights and in accordance with the laws of that Party; or

e) wherever polar bears have or might have been subject to taking by traditional means by its nationals.

2. The skins and other items of value resulting from taking under subparagraphs b) and c) of paragraph 1 of this Article shall not be available for commercial purposes.

ARTICLE IV

The use of aircraft and large motorized vessels for the purpose of taking polar bears shall be prohibited, except where the application of such prohibition would be inconsistent with domestic laws.

ARTICLE V

A Contracting Party shall prohibit the exportation from, the importation and delivery into, and traffic within, its territory of polar bears or any part or product thereof taken in violation of this Agreement.

ARTICLE VI

1. Each Contracting Party shall enact and enforce such legislation and other measures as may be necessary for the purpose of giving effect to this Agreement.

2. Nothing in this Agreement shall prevent a Contracting Party from maintaining or amending existing legislation or other measures or establishing new measures on the taking of polar bears so as to provide more stringent controls than those required under the provisions of the Agreement.

ARTICLE VII

The Contracting Parties shall conduct national research programmes on polar bears, particularly research relating to the conservation and management of the species. They shall as appropriate coordinate such research with the research carried out with other Parties on the management of migrating polar bear populations, and exchange information on research and management programmes, research results and data on bears taken.

ARTICLE VIII

Each Contracting Party shall take actions as appropriate to promote compliance with the provisions of this Agreement by nationals of States not party to this Agreement.

ARTICLE IX

The Contracting Parties shall continue to consult with one another with the object of giving further protection to polar bears.

ARTICLE X

1. This Agreement shall be open for signature at Oslo by the Governments of Canada, Denmark, Norway, the Union of Soviet Socialist Republics and the United States of America until 31st March 1974.

2. This Agreement shall be subject to ratification or approval by the signatory Governments. Instruments of ratification or approval shall be deposited with the Government of Norway as soon as possible.

3. This Agreement shall be open for accession by the Governments referred to in paragraph 1 of this Article. Instruments of accession shall be deposited with the Depository Government.

4. This Agreement shall enter into force ninety days after the deposit of the third instrument of ratification, approval or accession. Thereafter, it shall enter into force for a signatory or acceding Government on the date of deposit of its instrument of ratification, approval or accession.

5. This Agreement shall remain in force initially for a period of five years from its date of entry into force, and unless any Contracting Party during that period requests the termination of the Agreement at the end of that period, it shall continue in force thereafter.

6. On the request addressed to the Depository Government by any of the Governments referred to in paragraph 1 of this Article, consultations shall be conducted with a view to convening a meeting of representatives of the five Governments to consider the revision or amendments of this Agreement.

7. Any Party may denounce this Agreement by written notification to the

Depository Government at any time after five years from the date of entry into force of this Agreement. The denunciation shall take effect twelve months after the Depository Government has received this notification.
8. The Depository Government shall notify the Governments referred to in paragraph 1 of this Article of the deposit of instruments of ratification, approval or accession, for the entry into force of this Agreement and of the receipt of notifications of denunciation and any other communications from a Contracting Party specially provided for in this Agreement.
9. The original of this Agreement shall be deposited with the Government of Norway which shall deliver certified copies thereof to each of the Governments referred to in paragraph 1 of this Article.
10. The Depository Government shall transmit certified copies of this Agreement to the Secretary-General of the United Nations for registration and publication in accordance with Article 102 of the Charter of the United Nations.

(The Agreement came into effect in May 1976, three months after the third nation required to ratify did so in February 1976. All five nations ratified by 1978. After the initial period of five years, all five Contracting Parties met in Oslo, Norway, in January 1981, and unanimously reaffirmed the continuation of the Agreement.)

APPENDIX D

CONVENTION ON INTERNATIONAL TRADE IN ENDANGERED SPECIES OF WILD FAUNA AND FLORA (CITES) APPENDIX SYSTEM EXPLANATIONS

The Convention on International Trade in Endangered Species of Wild Fauna and Flora (CITES) is an international agreement regulating trade in a number of species of animals and plants, their parts and derivatives, and any articles made from them. Species are listed in one of three "Appendices," depending on how severely they are threatened by trade. Appendix I species are the most threatened. Once a species is listed in one of the CITES Appendices, all the nations that are parties to the convention (approximately 100 nations, including Canada) agree to the following conditions with respect to issuing permits and monitoring trade in these species:

APPENDIX I species are rare or endangered, and trade will not be permitted for primarily commercial purposes. Before trade is commenced, the importer must be in possession of a convention export permit issued by the government of the exporting nation and a CITES import permit issued by the government of the importing nation.

APPENDIX II species are not currently rare or endangered but could become so if trade is not regulated. The species being traded must be covered by an appropriate convention export permit issued by the government of the exporting nation before entry to or export from Canada will be allowed.

APPENDIX III species are not necessarily endangered but are managed within the listing nation. The species being traded must be covered by an appropriate convention export permit if trade is with the listing nation, or by a certificate of origin or a re-export certificate if trade is with a nation other than the listing nation, as required by the convention.

APPENDIX E

SAVING THE WILD CATS
IUCN CAT SPECIALIST GROUP

Cats have been part of the environment, culture, and mythology of human beings for thousands of years. The lion, in particular, has been widely used as a symbol of royalty and state to the present day. In pre-Columbian civilizations in Mexico and Central America, the jaguar had high ritual significance. The tiger has figured in the art and culture of the great civilizations of Asia. Domestic cats were revered in ancient Egypt, and in many countries today they rival the dog as a beloved companion of man.

Nevertheless, almost all species of wild cats are declining seriously in numbers because of human impact. Some subspecies are already extinct; others are on the brink of extinction.

Wild cats inhabit various parts of Africa, the Americas, and Eurasia. Lion, jaguar, leopard, tiger, and snow leopard are known as the big cats. Clouded leopard, cheetah, and puma [cougar] are also large in size, while the rest are smaller — African golden cat, Bornean bay cat, leopard cat, Chinese desert cat, caracal, jungle cat, pampas cat, Geoffroy's cat, kodkok, Iriomote cat, Andean cat, lynx, Pallas's cat, sand cat, marbled cat, black-footed cat, ocelot, flat-headed cat, rusty-spotted cat, bobcat, serval, wildcat (progenitor of the domestic cat), Asiatic golden cat, oncilla, fishing cat, margay, and jaguarundi.

The extinction of species of wild cats would be an inestimable loss to the world, not least because of their role as predators essential to natural ecology. It behooves us to make every effort to prevent such extinction because human activities are largely responsible for their deteriorating status.

WHY CATS SHOULD BE CONSERVED

Human beings have no right to eliminate other species. Indeed, in view of the extent of human domination of the natural environment, we have a responsibility and obligation to all species and to our descendants to perpetuate their existence. Extinction is forever.

The decline of a carnivore generally alters the ecological balance of its biological community. Cats are linked through predation to herbivores, which are, in turn, linked to each other through competition and to plant communities by their foraging. They are particularly sensitive to environmental disturbance, and the decline or disappearance of these vulnerable cat species serves as an indicator of changes in their ecosystem, which may be the result of natural phenomena or, as is increasingly the case in present times, of the impact of human activities. These changes frequently involve a deterioration in the human environment, such as the loss of forests and grasslands and their valuable animal and plant products, or impairment of water supplies essential to human life and agriculture. Furthermore, large cats, being at the pinnacle of the food chain, need considerable space, and are, therefore, key species in determining the area required to define an appropriate ecosystem.

In addition to the ecological consequences of the disappearance of these carnivores, many people feel a sense of inner loss when such magnificent and mysterious animals are gone from the wild.

PROBLEMS FACED BY THE CATS

Accelerating loss of habitat has now reached a critical stage as the human population continues to soar. In many cat ranges, remaining habitat represents but a small percentage of what existed in the past, and what remains could be wiped out in the near future.

Cats have long been hunted. They are killed because they have been viewed as competitors for prey. They are killed because they have taken livestock. They are killed for sport, and their body parts are used in some places as medicine. Young cats are captured for pets. And some, especially spotted cats, are killed for the fashion trade, which has often led to overexploitation.

At the same time, the disappearance of natural prey has frequently deprived cats of their normal sustenance and contributed to conflict with humans and their livestock, leading inevitably to reprisal killing of cats, often including those not actually involved.

Where cat populations have been reduced to small numbers they are increasingly vulnerable to extinction due to fortuitous local events, such as epidemics, fires, and floods. Some scientists also fear the possibility of deterioration through inbreeding depression and loss of genetic diversity in the long term, which might reduce the ability of small populations to adapt to changes in their environment.

THE DECLINE OF THE CATS

Cat populations have long been in decline, and today every indicator suggests that declines are accelerating and have reached, in some cases, a critical stage. The Asiatic lion is a classic example of decline because of human impact. Ranging 2,000 years ago from Asia Minor to Central India, it was hunted and exterminated, so that by the beginning of this century only a few survived in India's Gir forest. Fortunately, conservation efforts have succeeded in maintaining a lion population in the Gir, but it is confined to this single habitat, and thus is still dangerously vulnerable.

In 1947 the last recorded Asiatic cheetahs in the Indian subcontinent were shot. The subspecies still survives in Iran, but only in small numbers in fragmented habitat. The Bali tiger is thought to have already become extinct before 1940, and during this present decade of the 1980s, its neighbor, the Javan tiger, appears to have passed into oblivion. No trace of the Caspian tiger has been found for several decades, and reports suggest that the Amoy tiger, which is endemic to China, is on the verge of extinction, and that other subspecies of tiger may have vanished from the wild there by the end of the century.

The Indian or Bengal tiger had declined to dangerously low numbers by 1970, but has recovered as a result of dedicated, internationally supported conservation programmes implemented by the Indian and Nepalese governments. Nevertheless, it will remain vulnerable unless these programmes continue.

Among the small species, the Irimote cat, endemic to a Japanese island east of Taiwan, is nearly extinct because of destruction of its habitat and human overexploitation of its natural prey.

These examples of the decline of the cats and of suitable habitat are representative of the general situation throughout their world range.

PROBLEMS OF CAT CONSERVATION

There is still only limited knowledge of the distribution, numbers, biology, and behavior of almost all species of cat. Research to increase understanding of these factors is essential to the planning and implementation of effective conservation measures.

Economic planners and decision-makers often fail to recognize the importance for human welfare of wildlands, including ecosystems of which cats are part. Consequently, development programmes are carried out with little or no consideration of the longer-term impact, which may result in the decline and extinction of many species, including cats, as well as impoverishing the human environment.

As a result of increasing fragmentation of habitat and the pressure of human activities in their vicinity, large cats may become problem animals, patricularly through livestock predation, and in rare cases taking human life. Demands may then arise for elimination, not only of the offending animals, but of all the large cats in the area.

Insufficient resources are made available to pursue necessary research, and to implement protective measures and conservation management of natural habitats of cats, often because of failure to recognize their ecological significance and through lack of political will.

How Cats Can Be Conserved

Protected habitats of sufficient size and productivity to support viable populations of cats must be preserved, and linking corridors maintained wherever possible.

The distribution of each species and the habitat available to it needs to be established in detail, down to the level of discrete populations.

Legislation to insure long-term conservation of cat species and their prey, including controls on trade, national and international, must be passed and enforced.

Conservation of cats has to be reconciled with the needs of humans. Some conflict may be inevitable in areas where agriculture or livestock farming impinges on cat habitats, but it should be minimized by appropriate management measures. For many cats, and particularly large cats, parks and reserves may not be adequate. Land-use patterns in adjacent areas need to be designed so that they are compatible with use by both humans and cats.

Local people must feel that efforts are being made to protect their interests. Information about the role of cats and ways to conserve them should be part of conservation education at all ages and levels of the community, including the politicians, officials, industrialists, and businessmen who are the decision-makers.

Captive propagation programmes should be considered as an important precaution to serve as a genetic and demographic reservoir, which could, in appropriate circumstances, be used to reinforce wild populations.

All these measures should be included in an overall conservation strategy for each species to ensure its survival.

Conclusions

Species need not be lost provided action is taken to conserve them. Experience has shown that seemingly desperate situations can be reversed, if protection is given to species and their ecosystems.

The Cat Specialist Group is pledged to do all in its power to achieve the conservation of all cat species, and appeals for the cooperation of all people to insure that these magnificent animals continue to coexist with humans as they have through the ages.

APPENDIX F

DESCRIPTION OF PROPOSED
CARNIVORE CONSERVATION AREAS (CCAs)

THELON GAME SANCTUARY
(NORTHWEST TERRITORIES)

The Thelon Game Sanctuary is approximately 55,800 km², extending from the Back River to Dubawnt Lake. Wolverine, barren-ground grizzly, and wolves inhabit the area, as do a variety of other animals, including musk oxen, barren-ground caribou, red fox, peregrine falcons, gyrfalcons, loons, moose, and beaver. Although the sanctuary is over 160 km north of the treeline, relatively large spruce trees exist along the river and extend into the open tundra. Arctic willow, dwarf birch, labrador tea, lichens, and mosses are found in the sanctuary.

Little biological information is available on the large carnivores in the sanctuary, but they are known to be present in self-regulating populations. Therefore, more information is needed to precisely determine the importance of this sanctuary to these species. The Government of the Northwest Territories has already indicated that the current boundaries of the Thelon Game Sanctuary may not encompass minimum viable populations (MVPs) of various large carnivores there, particularly grizzly bears. General records of wildlife observations and reports indicate that grizzly bear densities decline in a west-to-east direction in the North. Western barren-ground grizzly densities range from 1 per 200–262 km², whereas in the East they may range as low as 1 per 400–500 km² (Lloyd, Personal Communication, 1990). Therefore, perhaps the current sanctuary would form the "core" of a CCA, with additional layers (or buffers) to provide adequate protection for enough individuals to encompass a MVP.

SOUTHERN MACKENZIE MOUNTAINS
(NORTHWEST TERRITORIES AND YUKON)

The second proposed CCA is situated in the southwest corner of the Northwest Territories and southeast corner of the Yukon in the Mackenzie Mountains. The area originally identified is bordered on the east by the Liard River, the west by the Hyland River in the Yukon, the south by the 60th parallel, and the northern boundary follows the northeast boundary of Nahanni National Park, then follows the eastern fork of the South Nahanni River northward to the Canol Heritage Road. Based on recent field work, WWF would support expanding this CCA to include the watershed of the Ram and North Nahanni rivers. The area is in excess of 65,000 km². The highest densities of grizzly bears in the Northwest Territories are found here, and most of it is closed to hunting. Wolverines and wolves inhabit the area, but exact numbers are not known. Wildlife is abundant and includes dall sheep, woodland caribou, moose, marten, lynx, gyrfalcon, and peregrine falcon. The area consists of rugged mountains and plateaus that exceed 2,000 m in elevation in certain areas.

Mixed and coniferous forests occur in the valleys and on the lower slopes. No permanent settlements or roads exist in the proposed CCA. Gold exploration is occurring near the confluence of Coal River and Rock River. Commercial forest areas do exist in the southeastern corner of the Yukon, extending through the river systems of La Biche, Beaver, Rock, Coal, and southern Hyland rivers (Dept. of Renewable Resources, Land Division, 1988). The city of Tungsten is located on the border of the CCA; therefore, adequate garbage disposal is needed to minimize problems with large carnivores, especially grizzly bears.

NORTH-CENTRAL YUKON

The third area identified is 60,000 km² and extends from the western border of the Yukon across to the eastern border of the Yukon and contains the Fishing Branch Game Reserve and the Peel River Game Reserve. The southern boundary has been defined as 65° 30' and the northern boundary defined by the 67th latitude. The only road through the area is the Dempster Highway, where hunting of grizzly bears has been closed along the highway for several years. Grizzly, wolverine, and wolf densities are good in this area, and the region is also a wintering area for the Porcupine Caribou Herd. Along the Fishing Branch River near Bear Cave Mountain, there is an excellent site for public viewing of bears in the September salmon runs. This area crosses the North Ogilvie Mountains ecoregion, the Eagle Plain ecoregion, the Peel River ecoregion, and the Coast Plain ecoregion.

Rocky Mountain Parks Complex
(Alberta and British Columbia)

The fourth area identified as a potential CCA is situated along the Alberta–British Columbia border. It includes the Rocky Mountain Parks complex and surrounding lands, an area approximately 50,000 km² in size. The area is bounded on the east by Forestry Trunk Road 940. The southern limit includes Kananaskis country and extends almost to Elkford, British Columbia. Included within the CCA are the Ghost River Wilderness, Siffleur Wilderness, White Goat Wilderness, and Willmore Wilderness areas. Also included are Banff, Jasper, Yoho, and Kootenay national parks. The western slopes of the Rocky Mountains include the following British Columbia provincial parks: Mount Assiniboine, Elk Lakes, Hamber, and Mount Robson, with a possible corridor to connect Wells Gray. Grizzly bears, black bears, wolves, cougars, wolverines, elk, moose, and deer inhabit the CCA. Though the area has many permanent human settlements, roads, and high recreational use, this CCA has the largest area of protective lands. WWF is currently funding work to encourage cross-jurisdictional cooperation in managing this area for the long-term conservation of large carnivores.

Northern British Columbia

The fifth area that has been identified as a potential CCA is the region between Mount Edziza Park and Atlin Park, in northwestern British Columbia, with a corridor connecting Mount Edziza Park to the Spatsizi Plateau Wilderness Park. The area is approximately 40,000 km² and is bounded by the Alaska border in the west. Grizzly bear and wolf densities in the area appear to be moderate to plentiful, and wolverines also inhabit the area, as do black bears, caribou, moose, and black-tailed deer. This CCA could be expanded northward to include the still wild Tatshenshini River Valley.

APPENDIX G

IUCN RESOLUTION ON WOLF/DOG HYBRIDS
(Passed April 20 & 24, 1990)

WHEREAS hybridization between wolves and dogs and the keeping of these hybrids as pets is becoming increasingly common in various countries, and especially in the United States and Canada;

WHEREAS most wolf-dog hybrids are poorly adapted to be pets, and there have been numerous fatal and non-fatal attacks on people and domestic animals which, in addition to being tragic and unavoidable, detract from the public perception of wild wolves; and

WHEREAS the perpetuation of these hybrids has no scientific or educational value; and whereas the ease with which escaped or unwanted hybrids can interbreed with wild wolves threatens the genetic integrity of wild populations;

NOW therefore be it resolved that the IUCN/SSC Wolf Specialist Group views the existence and expansion of wolf-dog hybrids as a threat to wolf conservation and recommends that governments and appropriate regulatory agencies prohibit or at least strictly regulate interbreeding between wolves and dogs and the keeping of these animals as pets.

CANADIAN WILDERNESS CHARTER

If you support the Canadian Wilderness Charter, and have not already signed on through the petition or charter card, please send a brief note to World Wildlife Fund Canada at the address indicated below, and make sure you sign your letter. Your name will then be recorded by WWF, and included as one of the one million Canadians we are seeking to sign the charter.

World Wildlife Fund Canada
Attention: Endangered Spaces Campaign
90 Eglinton Avenue East, Suite 504
Toronto, Ontario, M4P 2Z7
Tel: (416) 489-8800
Fax: (416) 489-3611

CANADIAN WILDERNESS CHARTER

1. Whereas humankind is but one of millions of species sharing planet Earth and whereas the future of the Earth is severely threatened by the activities of this single species,

2. Whereas our planet has already lost much of its former wilderness character, thereby endangering many species and ecosystems,

3. Whereas Canadians still have the opportunity to complete a network of protected areas representing the biological diversity of our country,

4. Whereas Canada's remaining wild places, be they land or water, merit protection for their inherent value,

5. Whereas the protection of wilderness also meets an intrinsic human need for spiritual rekindling and artistic inspiration,

6. Whereas Canada's once vast wilderness has deeply shaped the national identity and continues to profoundly influence how we view ourselves as Canadians,

7. Whereas Canada's aboriginal peoples hold deep and direct ties to wilderness areas throughout Canada and seek to maintain options for traditional wilderness use,

8. Whereas protected areas can serve a variety of purposes including:

a) preserving a genetic reservoir of wild plants and animals for future use and appreciation by citizens of Canada and the world,

b) producing economic benefits from environmentally sensitive tourism,

c) offering opportunities for research and environmental education,

9. Whereas the opportunity to complete a national network of protected areas must be grasped and acted upon during the next ten years, or be lost,

1. **We agree and urge:**
That governments, industries, environmental groups, and individual Canadians commit themselves to a national effort to establish at least one representative protected area in each of the natural regions of Canada by the year 2000,

2. That the total area thereby protected comprise at least 12% of the lands and waters of Canada as recommended in the World Commission on Environment and Development's report, *Our Common Future,*

3. That public and private agencies at international, national, provincial, territorial, and local levels rigorously monitor progress toward meeting these goals in Canada and ensure that they are fully achieved, and

4. That federal, provincial and territorial government conservation agencies on behalf of all Canadians develop action plans by 1990 for achieving these goals by the year 2000.

BIBLIOGRAPHY

Popular Articles and Books

Banci, V. 1986. The Wolverine in Yukon: Myths and Management. *Discovery,* December, 15(4):134-137.

Banci, V. 1987. Yukon wolverine: Research and Management. *International Trapper,* Winter 1987, Volume 2, Issue 1: 6-9.

Banfield, A.W.F. 1974. *The Mammals of Canada.* University of Toronto Press.

Brown, D.F. 1985. *The Grizzly in the Southwest: Documentary of an Extinction.* University of Oklahoma Press, Norman, Ok.

Brown, D.F. 1982. *The Wolf in the Southwest.* University of Arizona Press, Tucson, Az.

Brown, T. 1990. It's now or never. Editorial, *Outdoor Canada,* October 1990, page 6.

Burnett, J.A., C.T. Dauphiné Jr., S.H. McCrindle and T. Mosquin. 1989. *On The Brink: Endangered Species in Canada.* Western Producer Prairie Books, Saskatoon, Sask.

Cernetig, Miro. 1990. Who's afraid of Montana wolf? *Globe and Mail,* Oct. 15, 1990.

Chapman, J.A., and G.A. Feldhammer. 1982. *Wild Mammals of North America.* The Johns Hopkins University Press, Baltimore, Md.

DeBlieu, Jan. 1991. *Meant to be Wild: The Struggle to Save Endangered Species Through Captive Breeding.* Fulcrum, Golden, Colorado.

Dehr R. and R.M. Bazar. 1989. *Good Planets are Hard to Find!* Earth Beat Press, Vancouver, B.C.

Dekker, D. 1985. *Wild Hunters.* Canadian Wolf Defenders. Edmonton, Alta.

Fischer, H. 1988. Wolves for Yellowstone? *Defenders* March-April, 1988: 16-17.

Foreman, D., and H. Wolke. 1989. *The Big Outside.* Ned Ludd Books, Tucson, Arizona.

Glacier National Park Research Division. 1988. Wolves of Glacier (Pamphlet), Glacier National History Assoc., Montana, U.S.A.

Griss, P. 1990. *The Daily Planet: A Hands-On Guide to a Greener Environment.* Key Porter Books, Toronto, Ont.

Harding, A.R. 1909. *Wolf and Coyote Trapping.* A.R. Harding Publishing Co., Columbus, Ohio.

Herrero, S. 1985. *Bear Attacks: Their Causes and Avoidance.* Nick Lyons Books, Winchester Press, U.S.A.

Hoagland, Edward. 1976. *Red Wolves and Black Bears.* Random House, New York.

Horejsi, B.L. 1990. The grizzly bear management plan for Alberta: Fraud or folly? Speak up for Wildlife Foundation Research Report, Calgary, Alta.

Hornocker, M.G. 1969. Stalking the mountain lion—To save him. *Nat. Geog.* 136(5): 638-655.

Hummel, M. 1982. Taking aim at the wolf. *Nature Canada,* Spring, 1982, Canadian Nature Federation, Ottawa, Ont.

_____ . 1987. Do Carnivores have a future? Notes from an address to the Symposium on Bear-People Conflicts, April 6-10, 1987, Yellowknife, N.W.T.

————. 1989. Gen. ed. *Endangered Spaces*. Key Porter Books, Toronto, Ontario.

————. 1990. A conservation strategy for large carnivores in Canada. World Wildlife Fund Canada, Toronto, Ont.

Krueger, S. 1991. What would Darwin say? *Nature Canada,* Canadian Nature Federation, Winter, 1991: 13-17, Ottawa, Ont.

Larsen, T. 1984. We've saved the ice bear. *International Wildlife* 14:4-11.

Leopold, A. 1949. *A Sand County Almanac.* Oxford University Press (Ballantine Books edition, 1970, New York).

Littlejohn, B.M. and D.H. Pimlott. 1971. *Why Wilderness?* New Press, Toronto, Ont.

Livingston, J.A. 1981. *The Fallacy of Wildlife Conservation.* McClelland & Stewart, Toronto.

Mech, L.D. 1970. *The Wolf: The Ecology and Behavior of an Endangered Species.* Garden City, N.Y. American Museum of Natural History and Natural History Press.

————. 1988. *The Arctic Wolf: Living with the Pack.* Key Porter Books, Toronto, Ont.

McNamee, T. 1982. *The Grizzly Bear.* Alfred A. Knopf Inc., reprinted by McGraw-Hill, 1969.

————. 1986. Yellowstone's Missing Element. *Audubon,* January, 1986. 12-19. January.

Murray, John A. 1987. *Wildlife in Peril: The Endangered Mammals of Colorado.* Roberts Rinehart, Niwot, Co.

Parks Canada. 1979. *Parks Canada Policy,* Ottawa, Ont.

Pettigrew, S.L., ed. 1990. Livestock Guardian Dog. *Focus,* Spring, 1990, Northwest Wildlife Preservation Society, Vancouver, B.C.

BIBLIOGRAPHY

Pimlott, D.H. 1961. Wolf control in Canada. *Canadian Audubon Magazine*, Nov.-Dec., Ottawa, Ont., 1961.

Pollution Probe Foundation. 1989. *The Canadian Green Consumer Guide*. McClelland & Stewart, Toronto.

Russell, A. 1967. *Grizzly Country*. Alfred A. Knopf, New York.

Rutter, R.J. and D.H. Pimlott. 1968. *The World of the Wolf*. J.B. Lippincott Co., Philadelphia and New York.

Savage, C. 1988. *Wolves*. Douglas & McIntyre Ltd., Vancouver, B.C.

————. 1990. *Grizzly Bears*. Douglas & McIntyre Ltd., Vancouver, B.C.

Stirling, I. 1988. *Polar Bears*. University of Michigan Press, U.S.A.

Theberge, J.B., 1975. *Wolves and Wilderness*. Dent, Toronto, Ont.

Wright, B.S. 1972. *The Eastern Panther: A Question of Survival*. Clarke Irwin & Co., Toronto/Vancouver.

Young, F.M., and C. Beyers. 1980. *Man Meets Grizzly: Encounters in the Wild from Lewis and Clark to Modern Times*. Houghton Mifflin, Boston, Ma.

Young, S.P. 1946. *The Puma: Mysterious American Cat. Part I: History, Life Habits, Economic Status and Control*. Dover Publications Inc., New York.

Yukon Conservation Newsletter. 1984. Predator control: To kill or not to kill? April, 1984.

TECHNICAL PAPERS AND REPORTS

Alberta Forestry, Lands and Wildlife. 1989. Management plan for Grizzly bears in Alberta, Edmonton, Alta.

Amstrup, S.C. and O. Wiig, 1991. *Polar Bears*. Proceedings of the Tenth Working Meeting of the IUCN/SSC Polar Bear Specialist Group. Occasional paper of the IUCN Species Survival Commission, no. 7

Anderson, A.E. 1983. A critical review of literature on puma *(Felis concolor)*. Colorado Division of Wildlife Special Report No. 54.

B.C. Ministry of Environment. 1989. Wolf-prey dynamics and management. Proceedings of a Symposium held May 10-11, 1988, Wildlife Working Report No. WR-40.

Banci, V. 1987. Ecology and behavior of wolverine in Yukon. Unpublished M.Sc. thesis, Simon Fraser University.

———. 1991. The status of the grizzly bear in Canada in 1990. Committee on the Status of Endangered Wildlife in Canada, Ottawa.

Bath, A., H. Dueck and S. Herrero. 1988. Carnivore conservation areas. Draft Report, University of Calgary, Alta.

Beck, T., *et al.* 1982. South San Juan Mountains grizzly bear survey. Colorado Division of Wildlife Project SE-3-4. Endangered Wildlife Investigations. Colorado Division of Wildlife nongame files. Photocopy.

Bissell, S.J. 1980. Grizzly bear incident, September 1979 summary report. Compendium of reports, maps, photographs, correspondence, and newspaper columns. Colorado Division of Wildlife nongame files.

Bromley, M. 1988. The status of the barren-gound grizzly bear *(Ursus arctos horribilis)* in Canada. Northwest Territories Dept. of Renewable Resources, unpublished report.

Bunnell, F.L. 1978. Basic considerations for study and management programs: Constraints of small populations. *Threatened Deer*, IUCN, Morges, Switzerland, pp. 264-287.

Bunnell, F.L. and D.E.N. Tait. 1981. Population dynamics of bears—implications. *Dynamics of large mammal populations*, C.W. Fowler and T.D. Smith, ed. John Wiley and Sons, Inc., pp. 75-98.

———. 1985. Mortality rates of North American bears. *Arctic* 38 (4): 316-323.

Cahalane, V.H. 1939. The evolution of predator control policy in the national parks. J. Wild. Manage. 3: 229-237.

Canadian Wildlife Service. 1990. CITES Report: 1989 Annual Report for Canada, No. 17, Nov. 1990, Ottawa, Ont.

Carbyn, L.N., 1983. Management of non-endangered wolf populations in Canada. Acta Zool. Fennica. 174:239-43.

———, ed. 1983. *Wolves in Canada and Alaska,* Canadian Wildlife Service Report Series No. 45, Ottawa, Ont.

———. 1987. Gray and Red Wolf. Pages 359-376 in *Wild Furbearer Management and Conservation in North America.* M. Novak, J.A. Baker, M.E. Obbard and B. Malloch, eds. Ontario Trappers Assoc. and the Ontario Ministry of Natural Resources.

Carbyn, L.N., D. Meleshka, P.C. Paquet, S. McKinlay, D. Burles and J. Pesevdorfer. 1986. Ecological studies of wolves and coyotes in Riding Mountain National Park, Manitoba. Prog. Rep. 3. Canadian Wildlfe Service, Edmonton, Alta.

Carr, H.D. 1989. Distribution, numbers and mortality of grizzly bears in and about Kananaskis County, Alberta. Wildl. Research Ser. 3, Alta. For. Lands and Wildl., Fish & Wildl. Div., Edmoton, Alta.

Craighead John J., J.S. Sumner, and G.B. Skaggs. 1982. *A Definitive System for Analysis of Grizzly Bear Habitat and Other Wilderness Resources.* University of Montana Foundation, Missoula. Wildlife-Wildlands Monograph.

Dauphiné, C. 1987. Status report on the wolverine *(Gulo gulo)* in Canada. Prepared for the Canadian Scientific Authorities for CITES, CWS, Ottawa, Ont.

Day, G.L. 1981. The status and distribution of wolves in the Rocky Mountains of the United States. M.S. thesis, University of Montana, Missoula.

Defenders of Wildlife. 1989. Preserving communities and corridors. Gay Macintosh, ed. Defenders of Wildlife, Washington, D.C.

Dueck, H. 1990. Carnivore conservation and interagency cooperation: A proposal for the Canadian Rockies. Canadian Parks Service contract No. 1632-89-052 and World Wildlife Fund, Hull, Quebec.

Elton, C.S. 1927. *Animal Ecology.* Sedgewick and Jackson, London, England.

Elton, C.S. and Nicholson, M. 1942. The ten-year cycle in numbers of the lynx in Canada, *Journal of Animal Ecology,* 11:215-244.

Erickson, A.W. 1974. Evaluation of the suitability of the Gila Wilderness for re-establishment of the grizzly bear. Report to the U.S. Forest Service, Southwestern Regional Office, Contract 6-369-74. Typescript.

Federal/Provincial/Territorial CITES Committee. 1990. Status Report on black bear management in Canada concerning the need for a CITES listing, unpublished working document.

Forbes, G.J. 1990. A Discussion of predation and its role in ecological community functions. Unpublished Ph.D. comprehensive examination, Univ. of Waterloo, Ont.

Fritts, S.H. 1990. Management of wolves inside and outside Yellowstone National Park and possibilities for wolf management zones in the greater Yellowstone area. Pages 1-3 to 1-88 in Yellowstone National Park et. al. Wolves for Yellowstone? 586 p.

_____ . 1982. Wolf depredation on livestock in Minnesota. U.S.D.I. Fish and Wildlife Service Resource Publication 145. Washington, D.C.

Fuhr, B. and D.A. Demarchi. 1990. A methodology for grizzly bear habitat assessment in British Columbia. Ministry of Environment, B.C.

Garner, G.W., S.T. Knick and D.C. Douglas. 1990. Seasonal movements of adult female polar bears in the Bering and Chuckhi Seas. Int. Conf. Bear Res. and Manage. 8: (in press).

Ginsberg, J.R. and D.W. Macdonald. 1990. Foxes, wolves, jackals and dogs: Action plan for the conservation of canids. IUCN/SSC Canid Specialist Group, IUCN, Gland, Switzerland.

Gilpin, M.E. 1989. Population viability analysis. Endangered Species Update 6(10): 15-18.

Haber, G.C. 1988. Wildlife management in Northern British Columbia: Kechika-Muskwa wolf control and related issues. Wolf Haven America, Wash., D.C.

———. 1990. Exploration of wolf-ungulate systems in Alaska—a summary. Alaska Wolf Management Planning Team, unpublished.

Hash, Howard. 1987. Wolverine. Pages 574-598 in *Wild Furbearer Management and Conservation in North America*. M. Novak, J.A. Baker, M.E. Obbard and B. Malloch, eds. Ontario Trappers Assoc. and the Ontario Ministry of Natural Resources.

Hebert, D. 1988. The Status and Management of Cougar in British Columbia. Unpublished report, B.C. Ministry of Environment.

Hornocker, M.G. 1969. Winter territoriality in mountain lions. *J. Wildl. Manage.* 33:457-464.

IUCN-The World Conservation Union. 1990. *Species*. Newsletter of the Species Survival Commission, no. 15, December, 1990.

Jalkotzy, M. and I. Ross, 1989. Management plan for cougars in Alberta. Discussion draft, Alberta Forestry, Lands and Wildlife, Fish and Wildl. Div.

Jonkel, C.J. 1978. Black, brown (grizzly) and polar bears. Pages 222-248 in *Big Game of North America*, Stackpole Books, Harrisburg, Pa.

———. 1987. Brown Bear. Pages 465-474 in *Wild Furbearer Management and Conservation in North America*. M. Novak, J.A. Baker, M.E. Obbard and B. Malloch, eds. Ontario Trappers Assoc. and the Ontario Ministry of Natural Resources.

Kaminski, T. and J. Hansen. 1984. Wolves of central Idaho. Unpublished. Montana Cooperative Wildlife Unit, Missoula, Montana.

Keith, L.B. 1974. Some features of population dynamics in mammals. Proc. Inter. Cong. Game Biol., Stockholm 11:17-58.

———. 1983. Population dynamics of wolves. Pages 66-77 in *Wolves in Canada and Alaska: Their status, biology and management*. L.N. Carbyn, ed. Proc. Wolf Sympos. 1981. Can. Wildl. Serv. Rep. Ser. 45.

Kelsall, J.P. 1981. Status report on the wolverine *(Gulo gulo)* in Canada. Committee on the Status of Endangered Wildlife in Canada, Environment Canada, Ottawa.

Kolenosky, G. 1987. Polar Bear. Pages 474-485 in *Wild Furbearer Management and Conservation in North America.* M. Novak, J.A. Baker, M.E. Obbard and B. Malloch, eds. Ontario Trappers Assoc. and the Ontario Ministry of Natural Resources, Toronto, Ontario.

Kolenosky, G. and S.M. Strathearn. Black bear. Pages 442-455 in *Wild Furbearer Management and Conservation in North America.* M. Novak, J.A. Baker, M.E. Obbard and B. Malloch, eds. Ontario Trappers Assoc. and the Ontario Ministry of Natural Resources, Toronto, Ontario.

Larsen, T. 1986. Population biology of the polar bear *(Ursus maritimus)* in the Svalbard area. Norsk Polarinstitutt Skr., page 184.

Lavigne, D.M., A.L.A. Middleton, T.D. Nudds and U.G. Thomas. 1984. Wolf control programs in northeastern British Columbia. An assessment. Ad Hoc Committee of the Wildlife Society of Canada and Wildlife Biologists Section, Canadian Society of Zoologists, University of Guelph, Ont.

Lindzey, F. 1987. Mountain lion. Pages 656-668 in *Wild Furbearer Management and Conservation in North America.* M. Novak, J.A. Baker, M.E. Obbard and B. Malloch, eds. Ontario Trappers Assoc. and the Ontario Ministry of Natural Resources.

Lyster, S. 1985. *International Wildlife Law.* Cambridge, England, Crotius Publications.

Macey, A. 1979. Status report on grizzly bear *(Ursus arctos horribilis)* in Canada. Committee on the Status of Endangered Wildlife in Canada, Working paper, Ottawa, Ontario.

McBride, R.T. 1976. The status and ecology of the mountain lion of the Texas-Mexico border. M.Sc. Thesis, Sul Ross State University, Alpine, Texas.

McClure, J. 1990. S. 2674. A Bill to provide for the establishment of the gray wolf in Yellowstone National Park and the Central Idaho Wilderness Areas. 101st Congress. 2nd Session. May 22.

McCrory, W. and S. Herrero. 1987. Preservation and management of the grizzly bear in B.C. provincial parks, the urgent challenge. B.C. Parks and Outdoor Recreation Division, unpublished report.

McCrory, W., S. Herrero, G.W. Jones and E.D. Mallam. 1989. The role of the B.C. provincial park system in grizzly bear preservation. IBA Proceedings, Victoria, B.C.

McLellan, B.N. 1990. Relationships between human industrial activity and grizzly bears. *Int. Conf. Bear Res. and Manage.* 8:57-64.

McLellan, B.N. and D.M. Shackleton. 1989. Immediate reactions of grizzly bears to human activities. *Wildl. Soc. Bull.* 17:269-274.

Mech, L.D. 1991. Returning the Wolf to Yellowstone. Pages 309-322 in R.B. Keiter and M.S. Boyce, eds. The Greater Yellowstone Ecosystem. Yale University Press, New Haven.

Murie, A. 1971 (reprint of 1944 edition). The wolves of Mount McKinley. *Fauna of the National Parks of the United States*, Fauna Series No. 5.

Northwest Territories Wildlife Management Division. 1990. Discussion Paper towards the development of a Northwest Territories barren-ground grizzly bear management plan. Unpublished.

Peek, J.M., M.R. Pelton, H.D. Picton, J.W. Schoen, and P. Zager. 1987. Grizzly bear conservation and management: A review. *Wildl. Soc. Bull.* 15: 160-169.

Pimlott, D.H. 1967. Wolf predation and ungulate populations. *Amer. Zool.* 7:267-278.

Pimlott, D.H., J.A. Shannon and G.B. Kolenosky. 1969. The ecology of the timber wolf in Algonquin Provincial Park. Ont. Dept. Lands and Forest Res. Rep. (Wildlife) No. 87.

Ramsay, M.A. 1988. Reproductive biology and ecology of female polar bears *(Ursus maritimus)*. *Journ. of Zoology*, London, Ser. A. 214:601-34.

Reed, J.M., P.D. Doerr, and J.R. Walters. 1986. Determining minimum viable population sizes for birds and mammals. *Wild. Soc. Bull.* 14:255-261.

Ross, P.I. and M. Jalkotzy. 1988. The Sheep River cougar project—Phase II 1987-1988 progress report. Arc-Associated Resource Consultants, Ltd., Calgary, Alta.

Russell, R.H., J.W. Nolan, N.G. Woody and G.H. Anderson. 1979. Study of the grizzly bear in Jasper Park 1975-1978. Final report prepared for Parks Canada by the Canadian Wildlife Service, Edmonton, Alberta.

Schreiber, A., R. Wirth, M. Riffel and H. Van Rompney. 1989. Weasels, civets, mongooses and their relatives: An action for the conservation of mustelids and viverrids. IUCN. Gland, Switzerland.

Seidensticker, J.C.N., M.G. Hornocker, M.V. Wiles and J.P. Messick. 1973. Mountain lion organization in the Idaho Primitive Area. Wildl. Mangr. 35.

Servheen, C. 1990. The status and conservation of the bears of the world. Eighth International Conference of Bear Research and Management Monograph Series No. 2, International Association for Bear Research and Management.

————. 1990. Analysis of Asian trade in bears and bear parts. Research proposal of WWF U.S.

Sheeline, L., 1990. The North American black bear *(Ursus americanus):* A survey of management policies and population status in the U.S. and Canada, World Wildlife Fund U.S.

Sidorowicz, G.A. and F.F. Gilbert. 1981. The management of grizzly bears in the Yukon, Canada. Wild. Soc. Bull. 9(2):125-135.

Stirling, I. 1987. Status report on the polar bear *(Ursus maritimus).* Committee on the Status of Endangered Wildlife in Canada. Ottawa, Ont.

Taylor, M.K., D.P., DeMaster, F.L. Bunnell and R.E. Schweinsburg. 1987. Modelling the sustainable harvest of female polar bears. J. Wildl. Manage. 51:811-820.

Theberge, J.B. and D.A. Gauthier. 1985. Models of wolf-ungulate relationships: When is wolf control justified? Wildlife Soc. Bull. 13:449-458.

Tucker, P. 1988. Annotated gray wolf bibliography. U.S. Fish and Wildlife Service Region 6, Denver, Co 80225. 117 p.

Weaver, J. 1991. Landscape ecology of wolves in Jasper National Park. Progress Report, School of Forestry, University of Montana.

U.S. Fish and Wildlife Service. 1992. Recovery plan for the eastern timber wolf. Region 3, U.S.F.W.S. Twin Cities, Minnesota. 73 p.

U.S. Fish and Wildlife Service. 1989. Red wolf recovery plan. Region 2, U.S.F.W.S. Atlanta, Georgia. 110 p.

U.S. Fish and Wildlife Service. 1987. Northern Rocky Mountain wolf recovery Plan. Region 6, U.S.F.W.S. Denver, Colorado. 187 p.

Van Zyll de Zong, C.G. and E. Van Ingen. 1978. Satus report on the eastern cougar *(Felis concolor couguar)*. Committee on the Status of Endangered Wildlife in Canada, Ottawa, Ont.

INDEX